当代建筑师系列

都市实践
URBANUS

E6 空间　编著

中国建筑工业出版社

图书在版编目(CIP)数据

都市实践/E6空间编著. —北京：中国建筑工业出版社，2012.6
(当代建筑师系列)
ISBN 978-7-112-14252-1

Ⅰ. ①都… Ⅱ. ① E… Ⅲ. ①城市规划-建筑设计-研究
Ⅳ. ① TU984

中国版本图书馆 CIP 数据核字（2012）第 077029 号

整体策划：陆新之
责任编辑：刘 丹 徐 冉
责任设计：赵明霞
责任校对：肖 剑 陈晶晶

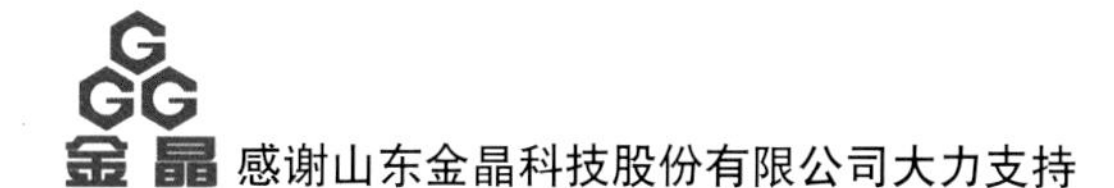
感谢山东金晶科技股份有限公司大力支持

当代建筑师系列
都市实践
E6 空间 编著
*
中国建筑工业出版社出版、发行（北京西郊百万庄）
各地新华书店、建筑书店经销
北京嘉泰利德公司制版
北京顺诚彩色印刷有限公司印刷
*
开本：965×1270 毫米 1/16 印张：$11\frac{1}{2}$ 字数：321 千字
2012 年 8 月第一版 2012 年 8 月第一次印刷
定价：98.00 元
ISBN 978-7-112-14252-1
(22327)

目　录　　Contents

都市实践印象

刘晓都 + 孟岩 + 王辉
文 / 黄元炤

刘晓都，1961 年出生，孟岩，1964 年出生，王辉，1967 年出生，三人皆毕业于清华大学建筑系，并先后赴美留学与工作，1999 年三人于美国纽约成立"URBANUS 都市国际设计"，2003 年更名为"URBANUS 都市实践"。"URBANUS"取意拉丁文的"城市"，以含义广泛且具有多意义的城市问题为他们的主要设计课题，致力于从广阔的城市视角和特定的城市体验中去解读建筑的内涵，同时追求现代主义的革命精神与理想，积极地介入社会，希望利用建筑去改善社会，让建筑使生活变得更加美好。

介入城市，是都市实践的一项潜模式。20 世纪 90 年代后，中国城市化进程日新月异，城市观念也非一成不变。建筑作为城市的构成元素，是影响城市的因素之一。所以，都市实践以介入城市、立足于城市作为其设计创作的出发点，有其定位的准确性。同时，他们注重的是设计，而不仅是市场与商业的需求，寻求城市中有意义的项目与课题。也由于都市实践回到国内实践时，深圳正进入大量建设完成的阶段，他们觉得可能要从意义或者质量上去发现城市中缺少了什么，于是他们发现深圳在城市空间的质量、尺度、内容和文化上似乎很欠缺，由此他们找到了切入点。

URBANUS 都市实践，一开始介入城市的潜模式是关注城市中待发展改建的空间，这些空间常被忽略。他们以介入城市空间去解决现实问题，回应中国当代城市中的复杂性。深圳地王城市公园第一期与第二期就是如此；这是两个城市开放空间改造的项目，都市实践在设计上以景观设计的手法，用坡道、路径整合场地周边原本各自独立的、零碎的、不完整的空地，形成一个新的、叠加的公共空间，启发人们不同的使用方式与体验，产生出新的城市公共活动场所，这是都市实践对城市公共空间的新的想法与诠释。在翠竹公园广场的设计中，他们利用补偿得来的空地做外部景观设计，采用了中式园林的庭院形式，围合出许多新的场域供民众使用，并形成中式园林的空间体验，建立自然与城市之间的中间生活地带。所以，可以看出都市实践关注于城市一开始是着重于景观与环境设计，并期望在城市中施以填充缝合的手法，整合出新的秩序与连接。另外，都市实践考虑问题的方式，是从问问题开始，先了解这个地方到底有什么问题，然后针对问题找一些解决的办法，这是一个相对客观的设计切入点。

城中村的改造，是都市实践关注的另一个城市课题。城中村是中国城市在 20 世纪末的高速发展下形成的灰色地带，城市规划无法管理，形成了垂直性发展的都市聚集地，也因无法管理而衍生出许多社会问题。当时对城中村的主导态度是拆除改造，以图改善当地的生活环境与品质。而都市实践从微观的城市视角发现城中村虽然看起来脏乱，但显得很有活力。他们觉得城市是需要互补的，不应该只存在干净，也应该容许脏乱。而城中村的脏乱带来了另一种城市边缘的生活模式，所以都市实践提出对于城中村的改造计划，以建筑学、社会学与城市文化学的探讨为前提，提出对当代城市生活的另一番新诠释。在深圳岗厦村中，都市实践与合作的研究者们提出不同于简单拆除重建的计划，他们承认这样自生自长的城市现象，接受底层劳动移民的人口，提出共存的说法；采取只拆除部分，填充、缝合以及加建的局部调整与修建的手法，通过密度调整，整合杂乱无章的建筑，维持了城中村特有的复杂活泼的既有结构。在深圳福新村中，都市实践提出城市中形成商业的策略，将原有人员迁出及功能置换，在改造建筑后形成高密度、混合的多功能商业村落。在深圳新洲中心，由于地处一片城中村，有着与周围城中村割裂的危险。于是都市实践提出将混合居住项目变成是城中村中的"村中城"，将住宅与商业混杂，创造出极大的兼容性，容纳各种城市活动，也刺激高度混合的有机社区的产生。而多样化的商业类型能够形成具有吸引力的购物区，创造出一个高度反差的环境。

可以观察出，都市实践在面对城市中特有的空间现象时，是一种切合现实的态度。即在非上层能主控的事实前，积极地承认，以一个直观的概念去面对城市。犹如现象学般以直观的方式，掌握事物的本质，同时排除掉所有预想与成见。呈现的内容可能推想出各种变化，而建筑态度上明显是一种以当地的观点介入设计，以城市的问题介入建筑，尊重底层的现实存在。深圳大芬美术馆，是都市实践积极与城市对话的代表性作品，他们意图在设计上与周边城市纹理与生活市井取得联系。都市实践提出"倒置的城中村"的概念，将地面层开放，引入街道集市及画廊商业行为，期望让各种艺术行为共存于使用的公共空间中，借以突破原有的层级位阶关系。此设计结合了周围城市的肌理，把美术馆、画廊、商业创作工作室等不同功能整合成一个整体，成为一个创新的艺术文化发源基地，并使艺术、生活、商业原本独立的三方面整合到一起。它可以汇集不同区域进入的人们，并形成相互交流、观摩与学习的艺术平台。所以，这个项目引入城市原本周边的商业形态，以突破原有的层级关系，仍然是以一种思考当地商业行为的观点介入建筑，积极地面对既有的环境。

土楼公舍，是都市实践朝向城市层面的多元统合的代表性作品。他们从社会议题层面着手，以传统文化为思考的立足点，将客家土楼提炼精化，并将人类居住单元与复合生活结构赋予了现代性与未来性。同时他们在设计中

Portrait

LIU Xiaodu + MENG Yan + WANG Hui

By Huang Yuanzhao

也切入普遍存在的民生议题，探讨在当今城市化过程中中低收入人群的居住问题，触及的是现代生活方式、邻里社群结构与社会价值等命题，在当代城市居住建筑中具有开拓意义。这个项目都市实践仍然维持着以社会学偏向于当地视点的方式思考着建筑。

都市实践也逐渐关注当代流行的建筑语言，在大连海中国美术馆，都市实践操作一种地景式建筑，他们以模拟地表起伏状态的连续性空间的创作，探讨了建筑与自然环境之间的关系，让环境的形态与意义进入建筑空间，形成连续性的空间状态。建筑形体经过一系列的转折、拉伸、撕裂、隆起与叠加等操作手法，构筑成一个地景建筑。同时都市实践也创造出一连串的走动过程，期望用这个走动过程让人与建筑、建筑与环境产生一种互动与交流的关系，进而让人们喜欢上这个建筑。

深圳招商海运大厦、深圳华美术馆、白云观珍宝花园这三个项目倾向于皮层性的建筑语言，但各自表述又不同。在深圳招商海运大厦中，都市实践将竖向建筑立面处理成类似集装箱的外在形式，穿插水平的线条带给人的立体感，倾向于皮层的表达。深圳华美术馆，在原有建筑外皮上加裹一层新的立面，由一个个大小不一、类似蜂窝状的六边形玻璃幕构成，并进行不同的类型组合，室内延伸立面上六边形的元素，利用六边形组合拉出一定进深，创造出展示的台面；或拉出一定进深之后再延伸折叠至顶棚，一直延续到对面的内墙面，形成新的室内整体蜂窝状的展示空间。这种新的立面蜂窝状表皮的延伸，使表皮的定义不再限于外部领域的新增或附加，而是作出表皮深度的新定义。表皮的形态已不再是立面上的平面叠加语法，而是立面加空间上的透视延伸。在白云观珍宝花园，都市实践关注到传统与现代的命题，用现代玻璃材料去体现古代文化，让现代与传统的东西并存，激活这个城市遗忘的角落与白云观本身存在的意义，让城市回归到当代，同时把一个好的东西，用当代语言表现出来，体现了都市实践做事的精神。

都市实践，以介入城市作为一项潜模式，以一种立足于城市层面、投入社会的设计创作为其始终，因为他们是继承了现代主义的革命精神与社会理想，冀望利用建筑试图去改变与改善社会，带有批判性，他们站在社会的角度积极去思考问题。10 多年来，都市实践未曾改变此现代主义精神的初衷，在此基点上关注到当代的问题，乃至于人。所以，他们认同平民，有很强的平民意识，关注平民，期望能够解决他们的建筑问题，仍是他们当前最重要的大事之一。

Liu Xiaodu (born in 1961), Meng Yan (born in 1964) and Wang Hui (born in 1967) all graduated from Qinghua University majoring in Architecture. After further study and work in America they collaboratively established and registered the company “URBANUS Design Worldwide” in the United States in 1999. Shortly afterwards the relocated studio in Shenzhen has revised its name to URBANUS Architecture and Design. The name “URBANUS” is derived from the Latin word for “Urban”, expressing the firm's primary design philosophy of searching for content in its architecture through the comprehensive reading of urban realities. The company's mission statement has clearly stated that URBANUS strives to “read architectural programs from the viewpoint of the urban environment in general, and ever-changing urban situations in particular.” URBANUS has also encouraged an active attitude in engaging in society with revolutionary design and passionate design attitudes, aiming to search for an alternative but utopian life with architectural design as an inter-medium.

URBANUS, following its mission statement's blueprint, tackles the urban issues from an approach named “urban engagement”. The drastic development of urbanization and modernization in China during the 1990s has relatively affected the fundamental perspective on urban planning, and consequently on design approaches in singular architecture entities. Architecture, as the most significant element from cities are constructed, has performed its role with vast influences on the mechanism and metabolism of the urban environment. Therefore, the “urban engagement” concept has become a major research topic in URBANUS, who continuously endeavors to solve critical and complicated problems in the urbanization process and to explore design possibilities emphasizing on urban characteristics rather than the knowingly conventional market-oriented or commercialized approaches. When URBANUS first relocated back to Shenzhen, the city was just entering the stage, in which massive constructions were completing. The company has hence decided to evaluate the already-planned city from a critical point of view, noting the absence of quality with diminishing cultural content, as well as the inappropriate scale and neglected ambiguous urban village issue.

The beginning of the "urban engagement" strategy has focused on inhabited ambiguous spaces in urban settings which are generally neglected in planning and development. URBANUS has proposed urban interventions to reflect complicated urban fabrics in modern Chinese cities. For example, in the Shenzhen Diwang Urban Park I and II projects the design attempts to renovate existing urban public spaces; by means of landscape design the urban surface system was reconfigured using terrace gardens, paved pedestrian paths and interconnected ramps. This is a unique explanation URBANUS has provided in translating a diverse urban public space, from which originally fragmented vacant lands emerged and were linked to reform intimate and user-friendly public spaces, encouraging social and community activities. Another instance in urban intervention is the Jade Bamboo Cultural Plaza, where the compensational lands were turned into a contemporary Chinese garden, providing zig-zagged terraced fields raised along the existing topography as a journey of the traditional courtyard experience. From the examples above, the urban strategies from URBANUS could be conceived of as pro-active and objective problem-solving processes, where the conflicts were initially addressed before the particular project on site, and urban voids were progressively filled and stitched in. As a result, a revitalized new urban public realm is emerging.

Another main urban research topic for URBANUS is the revitalization of the urban village. The unique phenomenon of "urban village" refers to the self-emerged grey zone on both the outskirts and downtown sections of major cities, commonly inhabited by transient and lower-income groups. Under China's extremely fast urbanization reform starting from the late 20th century, those urban villages are commonly associated with overcrowding, illegitimate vertical construction, un-hygienic living environments and other social problems. URBANUS has an alternative perspective on the nature of the villages, which are not regulated by any form of centralized urban planning. They are independent living models on the edge of major metropolises due to infinite modernization and increases in migrant population. From a micro-perspective, the city and the village should be complementary to each other; it is precisely a contradictory duality composing the full image of a particular urban living style. Starting from architectural studies, sociological research and urban-cultural analysis, URBANUS has proposed various hypothesis and public projects in re-habitation and regeneration of existing urban villages. In the project of Gangxia Village, the proposal acknowledges the existing social structure emerging from the village, and votes against the overall demolition program suggested by the city. URBANUS advises a revolutionary concept of respectful co-existence, and plans partial demolition and reconstruction with architectural intervention and infill, stitching and adjusting the existing heavily populated buildings. The density of the village is then re-evaluated, with maximum preservation of the dynamic and energetic nature of the village's mechanism. Another research example is rebuilding the Fuxin Village, in which the residents were relocated and the existing buildings were altered to perform a commercialization purposed function. The outcome was a high-density, mixed-used urban community with a hidden network infrastructure system, connecting existing buildings with variously proportioned passage rings, reinforcing the ordered but chaotic structure of the village. Last but not the least, in the planning of Xinzhou Central Plaza, the projects encountered problems stemming from the rupture and isolation of public spaces due to over-development of the surrounding urban villages. The complex is hence transformed as "city-in-village" zone amongst the urban village periphery. On one hand it aims to mix residential functions to contain various urban activities. On the other hand, through careful planning, the fusion between such diversities would stimulate the instinctive community with attractive divergence.

From the above-mentioned modes of research and design approaches, URBANUS has established realistic and objective strategies while investigating the spatial characteristics of an urban environment, using an intuitive phenomenological approach to actively extract the intrinsic quality of the facts, avoiding prejudice and pre-existing perceptions. In other words, the variation on architectural designs should arise from site-specific and socially-respectful attitude; materialize the urban problems into architecture interventions.

The dialogue between the company's design approaches and the city was best expressed in the Shenzhen Dafen Art Museum project. The design proposed a concept of "inverted urban village" which open up the ground level to connect the lively village town and the surrounding urban fabric. By introducing the casual street market and spaces for commercial galleries, the project hopes to diminish the existing negative hierarchical social structure, infiltrate and encourage creative activities in the new public spaces. The museum has combed through the periphery of the village's context, and integrated the independent three social groups–artists, residents and businessmen–to enable a forum or communication platform for interaction, observation and education. The project has further evaluated an urban intervention approaches based on the analysis of local commercial pattern through a series of conversation between existing environment and hidden hierarchy.

Tulou Collective Housing project is a representational example of URBANUS' hypothesis research towards diversification and integration of multiple urban layers. Commencing from a commonly noticed social problem, the Tulou Collective Housing's essence was extracted based on a traditional cultural phenomenon. It provided in-depth discussion of variable metamorphoses on residential typology and possible micro-urbanization possibilities, as well as tackled the complexity and conflicts in the living mechanism within the low-income group under modernization. The logic and process of the tulou program has set up a solid foundation and excellent precedent between translating research-based feasibility studies and design realization. It also reflects part of URBANUS statement, namely to evaluate architectural concepts potential based on site-specific sociological orders.

In addition to theoretical urban research, URBANUS has in recent project gradually turned its attention to contemporary elements in architecture vocabularies. Practicing in a land-art architectural experiment, the Dalian Maritime Art Museum implies a continuous space experience of the movement of the topography. Through actions of folding, tearing, extruding and repeating, the architecture was explicitly expressed as a sequence of motions. The fluent movement within the spaces, the building and the site has extended interactive possibilities.

Some other examples of alternative architectural vocabularies include Shenzhen Merchant Maritime & Logistics Tower, Shenzhen OCT Art & Design Gallery and White Cloud Temple Jewelry Garden. The three projects all indicate a façade-focused language while expressing distinct characteristics. The vertical façade of the Shenzhen Merchant Maritime & Logistics Tower was scarred and sliced in a three-dimensional imitation of a shipping container. In the OCT Art & Design Gallery, the main architectural gesture was to wrap the entire warehouse with a hexagonal glass curtain wall. With differentiated arrangement of four different sizes of hexagons the honeycombed structure is further extruded towards the interior spaces, extending, stretching and unfolding until it connects to the ceiling of the exhibition space. The unique inter-dependent façade and interior system have re-defined the existence of the skin of the building; it becomes more than an additional 2-Dimensional extension of a renovation program, but more importantly it allows the façade to intimately relate back to the spatial unity of the building. Lastly, in the White Cloud Temple Jewelry Garden, URBANUS critically considered the proposition between modernity and tradition. To pinpoint the neglected urban void in the temple the project used modern glazing material to manifest traditional cultural patterns, activating the site by returning its essence to the contemporary urban environment.

URBANUS, by believing in realism and holding a vision of revolutionary social ideology, strives to build its reputation as a design firm through interventions based on urban reality. In the past decade, the firm has persistently scrutinized critical design in architecture with social responsibility, and anticipated the progressive development the community. URBANUS firmly stresses contemporariness as its original intention in its mission statement. With a consciousness toward common citizens and vulnerable groups, they will continuously pinpoint and evaluate the architectural problems in society.

土楼公舍　广东 南海

Tulou Collective Housing, Nanhai, Guangdong

设计时间 / Design：2005 ~ 2007
建成时间 / Completion：2008
用地面积 / Site Area：9141m²
建筑面积 / Building Area：13711m²
项目组 / Design Team：刘晓都 孟岩 | 朱加林 | 李达 尹毓俊 | 黄志毅
李晖 程昀 黄煦 左雷 丁钰 魏志姣 黎靖 王雅娟 郑岩 沈艳丹
合作 / Collaborators：郭群设计（室内） 黄扬设计（标识）
博万建筑设计（施工图）
业主 / Client：深圳万科房地产有限公司
摄影 / Photographer：杨超英

土楼是客家民居独有的建筑形式。它是用集合住宅的方式，将居住、贮藏、商店、集市、祭祀、公共娱乐等功能集中于一个建筑体量，具有巨大凝聚力。将土楼作为当前解决低收入群体的住宅问题的方法，不只是形式上的承袭。土楼和现代宿舍建筑类似，但又具有现代走廊式宿舍所缺少的亲和力，有助于保持低收入社区中的邻里感。将"新土楼"植入当代城市的典型地段，与城市空地、绿地、立交桥、高速公路、社区等等典型地段拼贴，这些试验都是在探讨如何用土楼这种建筑类型去消化城市高速发展过程中遗留下来的不便使用的闲置土地。由于获得这些土地的成本极低，从而使低收入群体的住宅开发成为可能。土楼外部的封闭性可将周边较差的环境予以屏蔽，内部具有向心性同时又创造出温馨的小社会。将传统客家土楼的居住文化与低收入群体的住宅结合在一起，更标志着低收入人群的居住状况开始进入大众的视野。

这项研究的特点是分析角度的全面性和从理论到实践的延续性。对土楼原型进行尺度、空间模式、功能等方面的演绎，然后加入经济、自然等多种城市环境要素，在多种要素的碰撞之中寻找各种可能的平衡，这种全面演绎保证了丰富经验的获得，并为深入的思考提供平台。从调查土楼的现状开始，研究传统客家土楼在现代生活方式下的适应性，将其城市性发掘出来，然后具体深化，进行虚拟设计，论证项目的可行性，最终将研究成果予以推广，这样从理论到实践的连续性研究，是"新土楼"构想的现实性和可操作性的完美结合。

Tulou is a dwelling type unique to the the Hakka people. It is a communal residence between the city and the countryside, integrating living, storage, shopping, religion, and public entertainment into one single building entity. Traditional units in tulou are evenly laid out along its perimeter, like modern slab-style dormitory buildings, but with greater opportunities for social interaction. By introducing a "new tulou" to modern cities and by carefully experimenting its form and economy, one can transcend the conventional modular dwelling into urban design. Our experiments explored ways to stitch the tulou within the existing urban fabric, which includes green areas, overpasses, expressways, and residual land left over by urbanization. The cost of residual sites is low due to incentives provided by the government; this is an important factor for the development of affordable housing. The close proximity of each tulou building helps insulate the users from the chaos and noise of the outside environment, while creating an intimate and comfortable environment inside.Integrating the living culture of traditional Hakka tulou buildings with affordable housing is not only an academic issue, but also implies a more important yet realistic social phenomenon. The living conditions of impoverished people is now gaining more public attention. The research of tulou dwelling is characterized by comprehensive analyses ranging from theoretical hypothesis to practical experimentation. The study examined the size, space patterns, and functions of tulou. The new programs also inject new urban elements to the traditional style, while balancing the tension between these two paradigms. As a consequence of such comprehensive research, the tulou project has accumulated layers of experiences in various aspects. The project provided a platform for an in-depth discussion on feasibilities and possibilities of contextualizing the variable metamorphoses of traditional dwelling modules with an urban reality. It also introduced a series of publications and forums on future hypothetical designs for a "new tulou project". The logic and design process of the tulou program set up a solid foundation and excellent precedent for translating research-based feasibility studies to design realization.

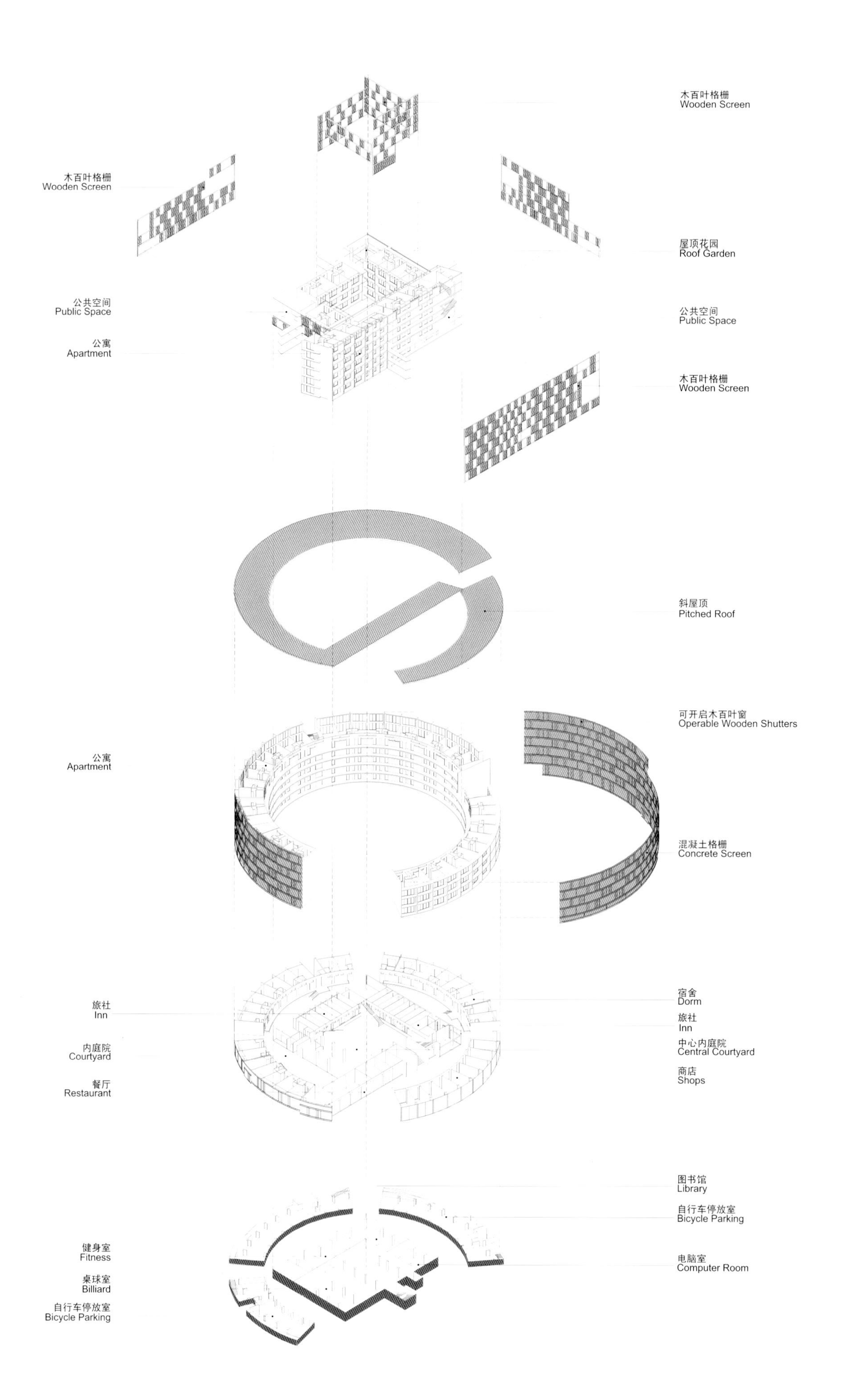
木百叶格栅
Wooden Screen
木百叶格栅
Wooden Screen
屋顶花园
Roof Garden
公共空间
Public Space
公共空间
Public Space
公寓
Apartment
木百叶格栅
Wooden Screen
斜屋顶
Pitched Roof
可开启木百叶窗
Operable Wooden Shutters
公寓
Apartment
混凝土格栅
Concrete Screen
旅社
Inn
宿舍
Dorm
旅社
Inn
内庭院
Courtyard
中心内庭院
Central Courtyard
餐厅
Restaurant
商店
Shops
图书馆
Library
自行车停放室
Bicycle Parking
健身室
Fitness
电脑室
Computer Room
桌球室
Billiard
自行车停放室
Bicycle Parking

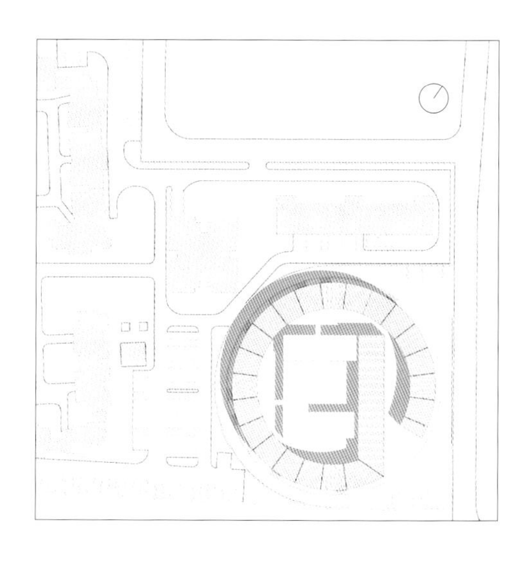

总平面图 / Site Plan

密度研究 / Density Study

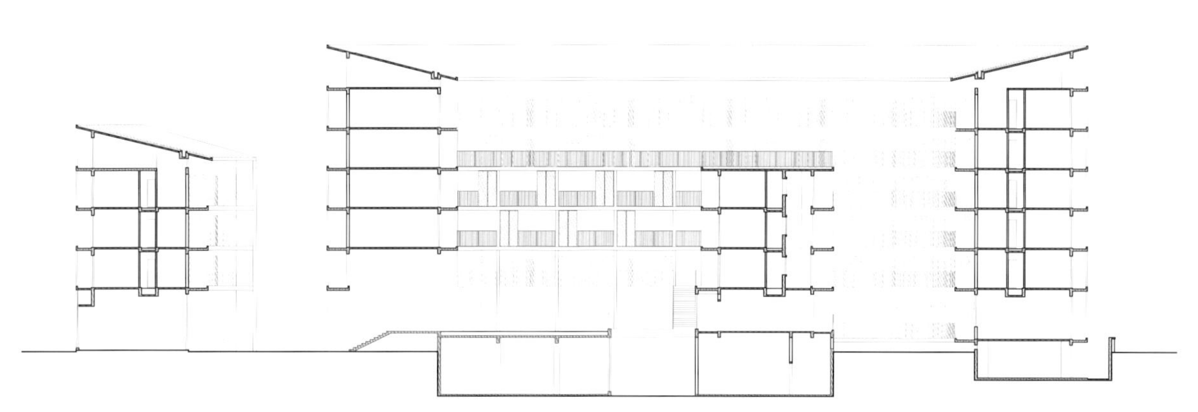

东西向剖面图 / Section East-West

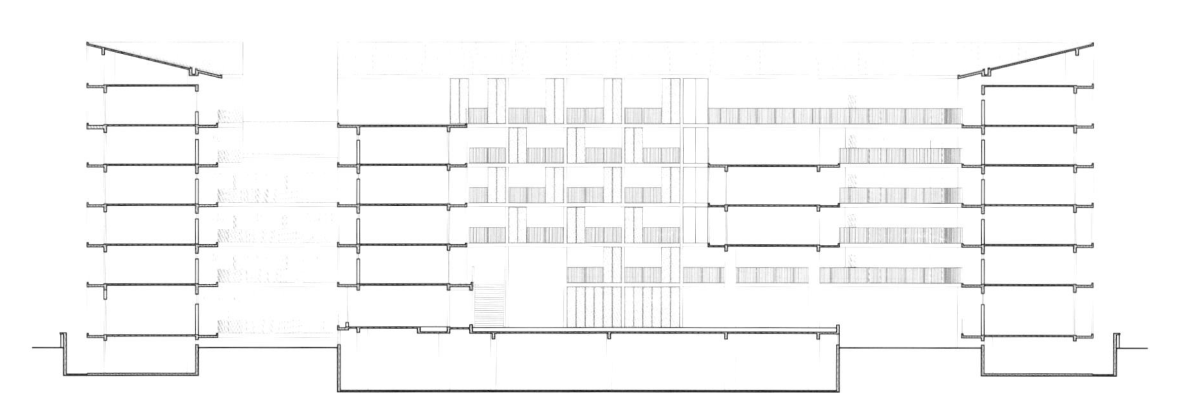

南北向剖面图 / Section North-South

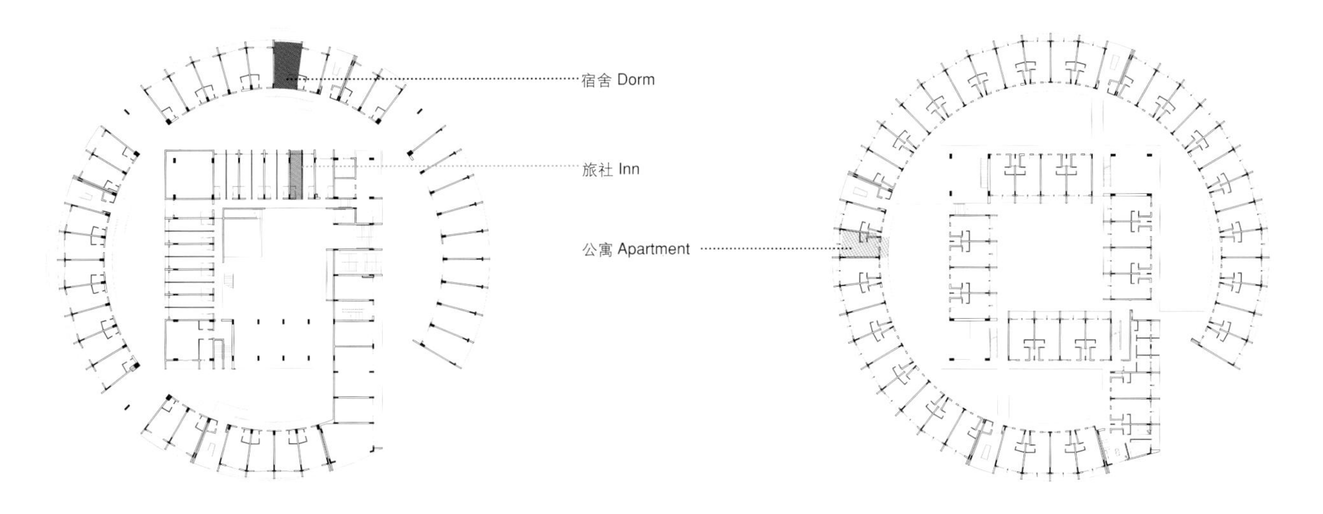

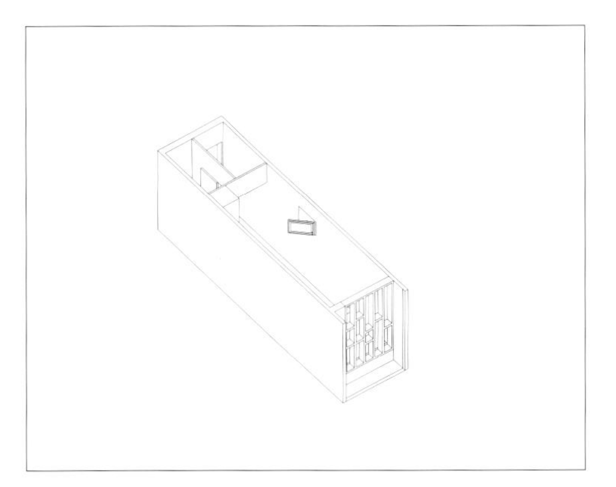

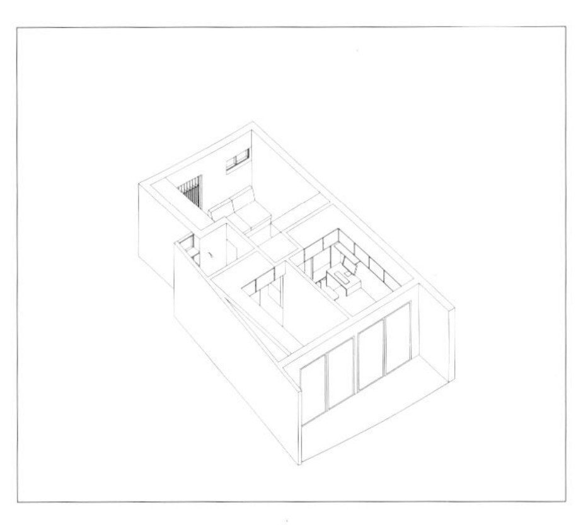

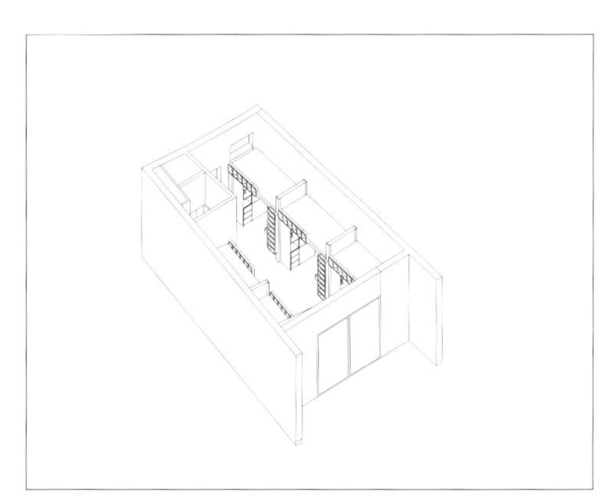

户型 / Unit Types

政治中心
Political Center

城市边缘
Edge of City

公园
Park

高档小区
Bourgeois Residence

商业中心
Commercial Area

立交桥
Overpass

城中村
Urban Village

公路边绿化带
Green Belt
Beside Highway

工厂区
Industrial Area

深港边界
Border of SZ and HK

体育中心
Sports Center

滨水保留地
River Bank

废弃停车场
Abandoned Parking Lot

美伦公寓＋酒店 深圳

Maillen Hotel & Apartment, Shenzhen

设计时间 / Design：2005 ~ 2009
建成时间 / Completion：2011
用地面积 / Site Area：13198m^2
建筑面积 / Building Area：21540m^2
项目组 / Design Team：孟岩 | 朱加林 | 黄志毅 姚晓微 郑颖 | 邢果
魏志姣 | 左雷 刘小强 叶沛军 李达 沈艳丹 嵇羽宇 张震 刘浏
夏淼 熊慧晶 | 丁钰 Cedric Yu 黎靖 黄艺宏 廖志雄
合作 / Collaborators：水平线空间环境设计（室内）
广东省建筑设计研究院深圳分院（施工图）
业主 / Client：深圳市招商房地产有限公司
摄影 / Photographer：吴其伟

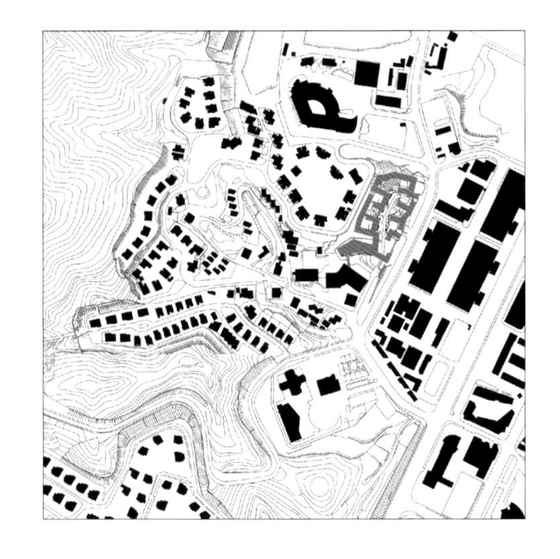

当前中国城市中的住宅生产几乎已被低市场风险的公式化手法彻底统治，尤其是商品房，几乎没有留给建筑师多余的想象空间。作为以公寓和主题式酒店为主体的美伦公寓，是以出租为主，因此能够摆脱陈词滥调的销售说辞，而去探索诗意的居住理念。

该项目坐落于深圳市蛇口半山别墅区脚下，典型的山丘地形激发了设计意向：山外山，园中园。中国人通常使用"山—水"和"园—林"来表达一种对生活的理解和对自然的向往。运用此概念，设计师希望能塑造全新的居住空间，将传统的居住模式和现代生活结合。

依地势和空间的围合要求，盘旋而出一段山形般波折起伏的建筑形体，把基地环抱其中，实现"山外青山楼外楼"的空间意境。园林与建筑结合是我国传统建筑的基本组合方式。总体上用建筑围合成了一个大园子，园子中凿咫尺小池为镜，以桥为舟，一个个房子从"建筑山"生长出来，临水而居，一幅江南水乡景色。当然仅此仍嫌不足，在大园中又勾画出五个相互连通的小庭院，各具特色。在基地中穿行，移步异景，桃花源的意境悄然地嫁接到现代城市的生活当中。

Current residential designs in many Chinese cities are dominated by low-risk formulated plans and generic marketing strategies. This sales-dependent development has left few possibilities for architects to explore further creativity in design and experimentation.

This hillside apartment, on the other hand, has its main goal on renting rather than sales, focusing on service apartments and a garden-themed hotel. Therefore the concept in designing and planning is able to depart from the stereotypical categories, seeking a more poetic and dynamic way of residential design.

The project returns to fundamental ideas in Chinese living as expressed by the saying "hills outside hills, and gardens inside gardens", an idea referring to a continuous and occasionally repeating rhythm of space and form found in many traditional villages and mountainous landscapes. The relationship between nature and buildings is blurred in an attempt to create a new generation of urban living.

Located on the foot of the "south mountain", the site is terraced and sloped. The buildings gently grow out from the landscape, taking on the angular characteristic of the geography while offering ponds and courtyards to the residents. Views from the units extend to several smaller courtyards where bamboo, pine, and plum blossom can be found. In the center of the site, a modest walkway forms a link over the water, bridging the interconnected gardens.

景观总平面 / Landscape plan

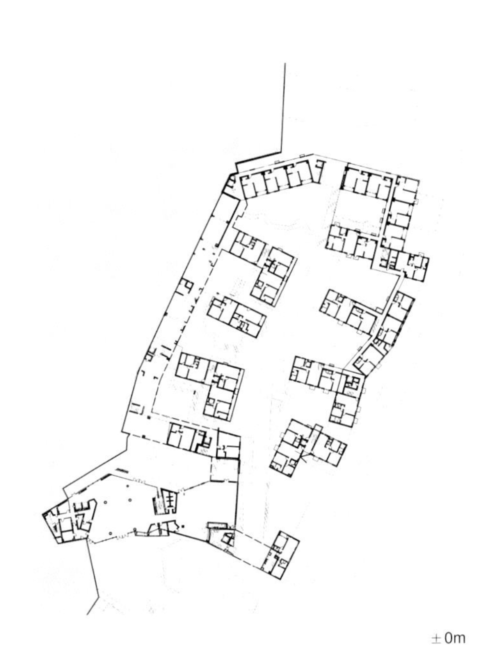

±0m

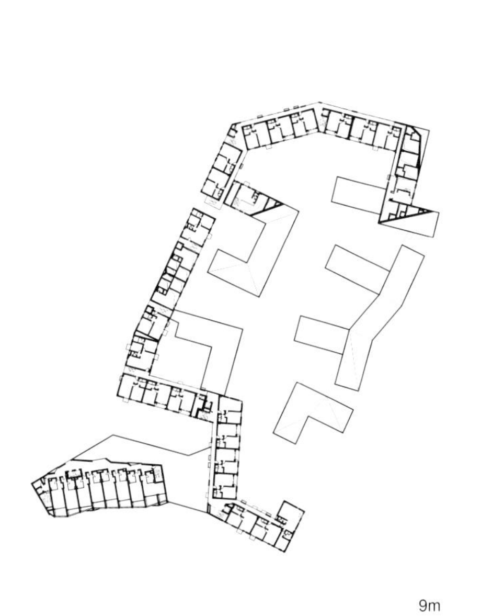

9m

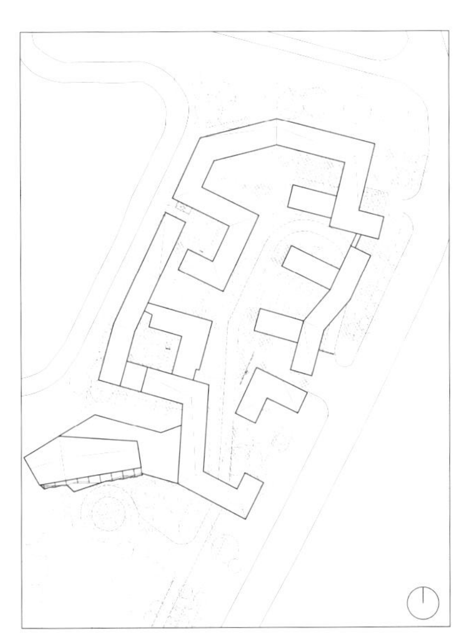

屋顶平面图 / Roof Plan

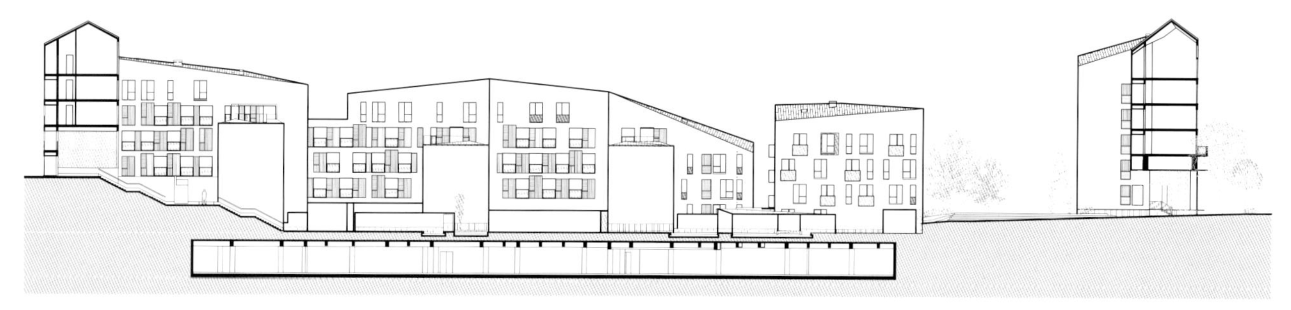

e-e 剖面图

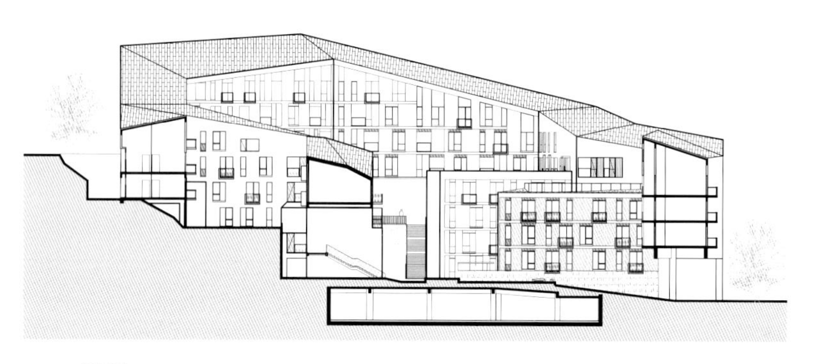

g-g 剖面图

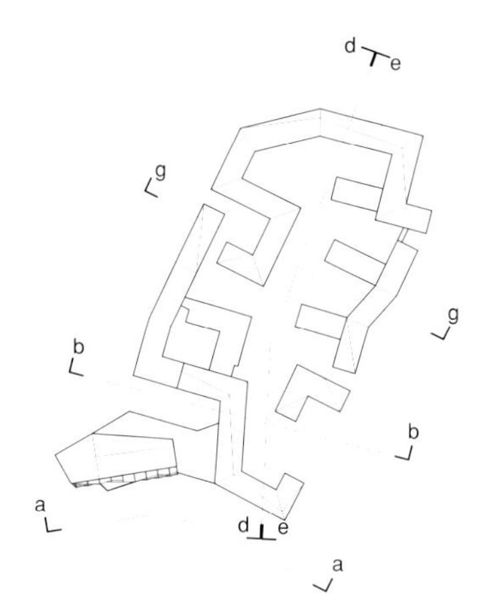

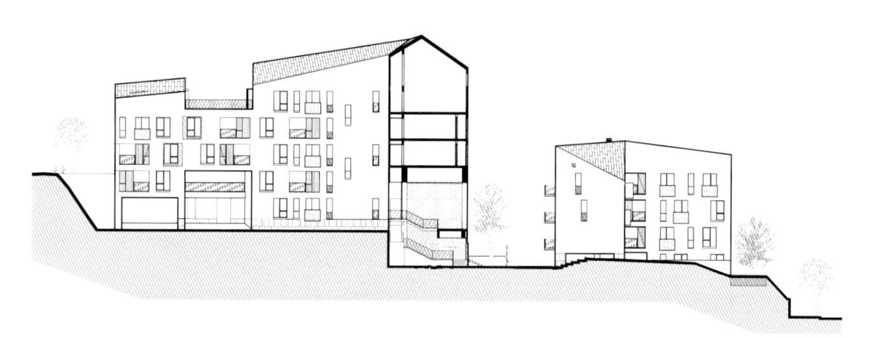

b-b 剖面图

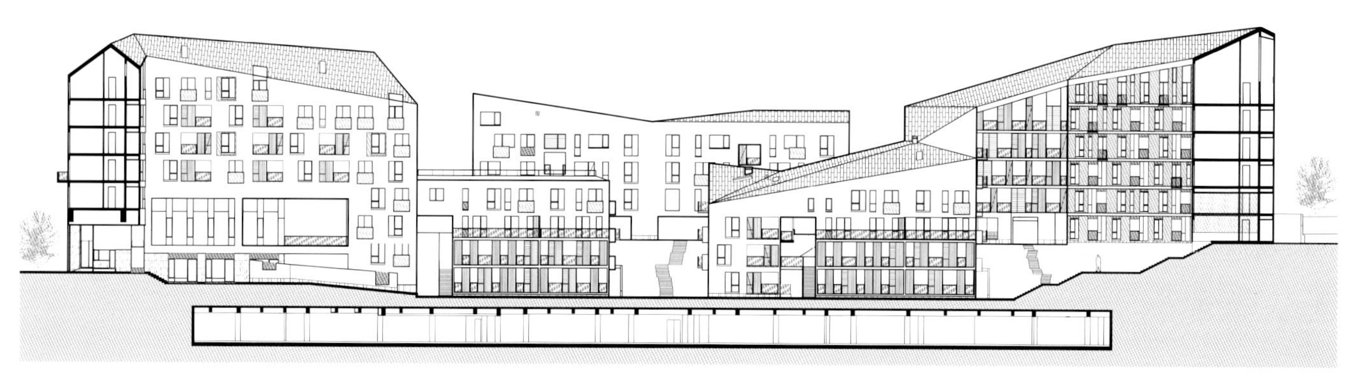

d-d 剖面图

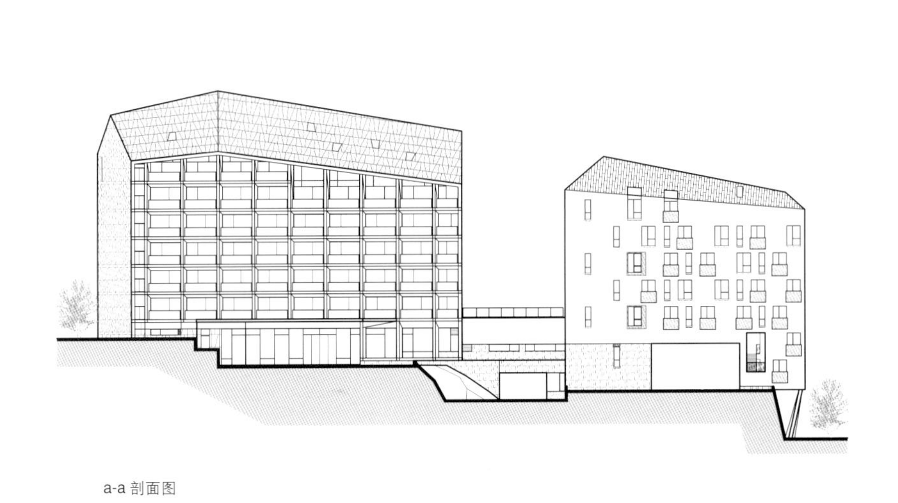

a-a 剖面图

华·美术馆 深圳

OCT Art & Design Gallery, Shenzhen

设计时间 / Design：2006 ~ 2008
建成时间 / Completion：2008
用地面积 / Site Area：2700m²
建筑面积 / Building Area：2620m²
项目组 / Design Team：孟岩 刘晓都 | 朱加林 吴文一 | 邓丹 姚晓微 | 程昀 黎靖 魏志姣 Cedric Yu
合作 / Collaborators：郭群设计（室内） 坊城建筑设计（施工图） 珠海晶艺玻璃（幕墙） 黄扬设计（平面）
业主 / Client：深圳市华侨城酒店集团公司
建筑摄影 / Photographer：强晋 孟岩

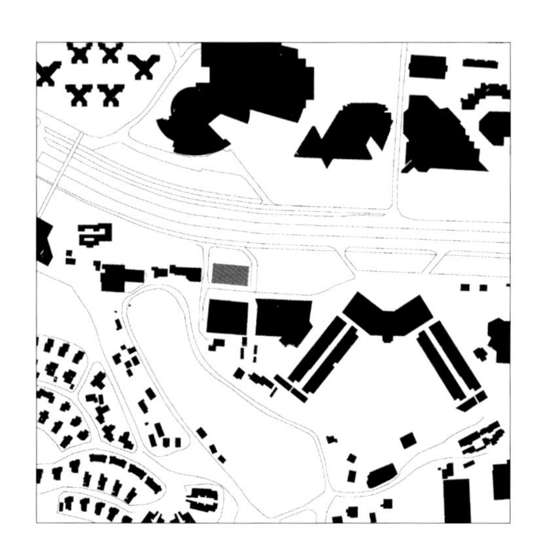

位于深圳深南大道南侧的“华·美术馆”，原是建于20世纪80年代早期的深圳湾大酒店的洗衣房。在高速发展的城市中，这座存在于西班牙式的华侨城洲际酒店和典雅的何香凝美术馆夹缝中的旧厂房，因其单调的建筑形式早已成为不为人留意的都市残留物。厂房的产权方考虑到其优越的地理位置，决定将其保留并改造为艺术展馆。

虽是邻近的国家级美术馆展览空间的延伸，但酒店展馆的特殊定位，决定了改造后展馆的独特性：其设计既要突显个性，与两边建筑风格形成差异性对比，同时也要体现与两端建筑的关系及整体性。改造策略完整地保留了原建筑立面的窗墙体系，加建的立面通过包裹的手法，将单一的原始六边形通过复杂有机的组合形成由实至虚、由小到大和多层次渐变的三维视觉效果。从而在车辆由西至东快速通过的瞬间，形成强烈的视觉冲击力，通过立面结构的缩小放大，逐层递减，如同面纱般轻轻揭开，最终透出原建筑立面的戏剧性变化过程。

展馆的室内设计再次运用立面所含有的六边形元素作为基本平面形态，在竖向上作90°的拉伸，形成一系列折叠平面互相交叉、互相切入的、复杂但带着明确功能元素的公共空间。这种表达形式改变了原本单调的立面几何图案，用三维方式生成新的室内空间，这种“突变”形式使设计产生了不可预料的惊喜结果。这个旧洗衣房的华丽转身，也折射了深圳这个城市从早期的工业城市向当代国际大都市的转变。

The site has had a rather unremarkable history. Originally constructed as laundry facility for Shenzhen Bay Hotel in the early 1980's, it is situated along the main road, between a Spanish-style OCT Hotel and the Hexiangning Art Museum. Over many years, the warehouse itself remained unaltered while the city around it rapidly transformed. Considering the significance of its location, the owner has decided to preserve the existing building and remodel the warehouse into an art exhibition loft adjacent to the main gallery. For URBANUS, the remodelling of the site poses difficult questions of how to address the existing urban condition, and how new interventions would relate to it. The outcome of the remodeling would be more than just an extension wing of the main gallery. It would be a strong landmarking piece connecting two distinguished existing buildings, yet still retain an unmistakably independent character of its own. The main architectural gesture is to wrap the entire warehouse with a hexagonal glass curtain wall. The pattern is created from 4 different sizes of hexagons. As a result, the new wall becomes a lively theatrical screen. Viewing from the mainroad in fast-moving traffic, the façade has an animatedly strong visual impact, like un-veiling layers of embroidered Chiffon.

The geometric pattern is more than just surface deep. It is actually a three-dimensional matrix of intersecting elements that project into the gallery spaces, structuring the building's interior design. The various depth of extrusion of the simple hexagon perimeters has defined the internal public and private spaces. The result is the creation of delightful and unexpected spatial experiences. The transformation of the former laundry factory building has been recorded in the significant pages of history, where Shenzhen has quietly transcended itself from an industrialization-based city into a international metropolis.

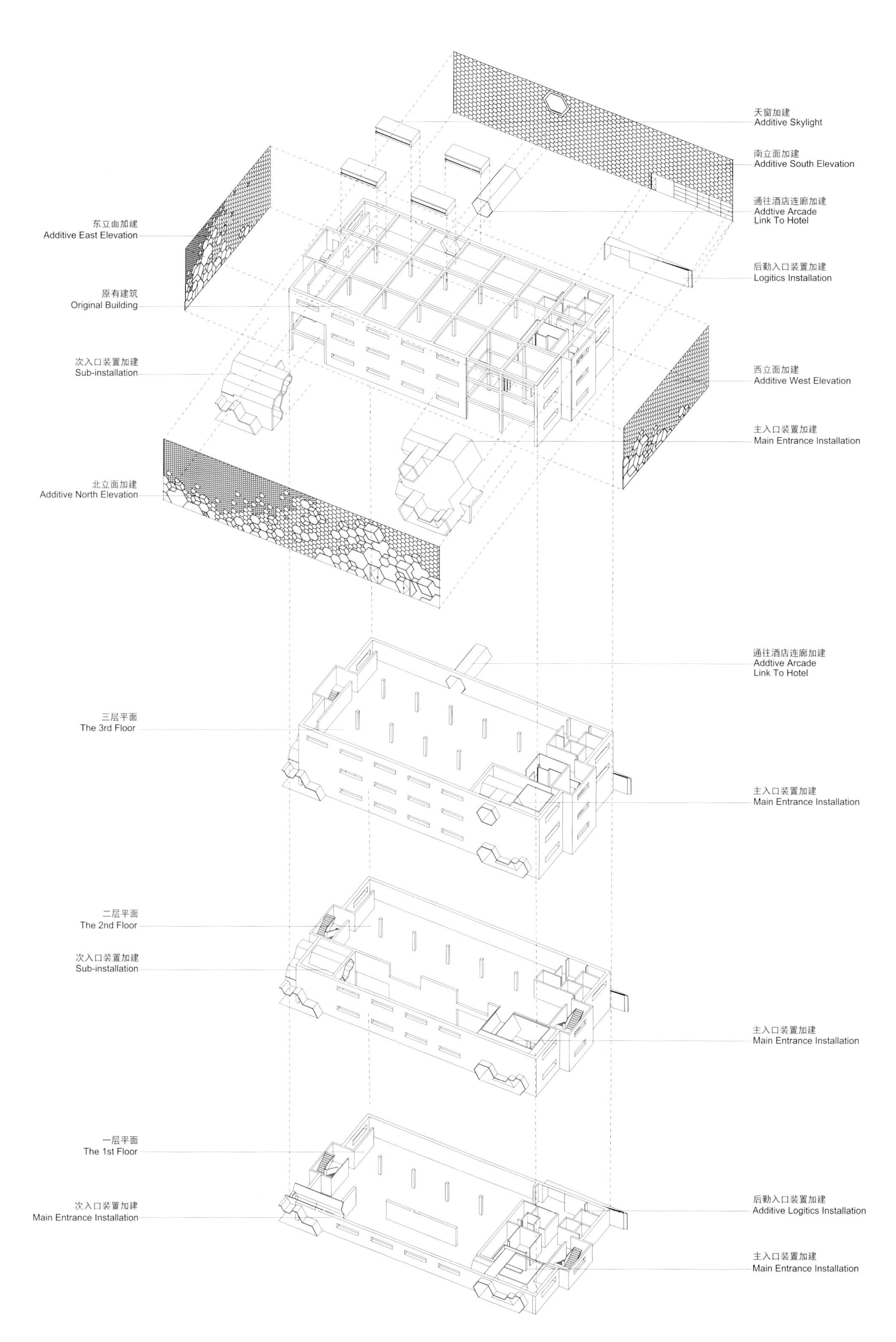
天窗加建
Additive Skylight
南立面加建
Additive South Elevation
通往酒店连廊加建
Addtive Arcade
Link To Hotel
东立面加建
Additive East Elevation
后勤入口装置加建
Logitics Installation
原有建筑
Original Building
次入口装置加建
Sub-installation
西立面加建
Additive West Elevation
主入口装置加建
Main Entrance Installation
北立面加建
Additive North Elevation
通往酒店连廊加建
Addtive Arcade
Link To Hotel
三层平面
The 3rd Floor
主入口装置加建
Main Entrance Installation
二层平面
The 2nd Floor
次入口装置加建
Sub-installation
主入口装置加建
Main Entrance Installation
一层平面
The 1st Floor
次入口装置加建
Main Entrance Installation
后勤入口装置加建
Additive Logitics Installation
主入口装置加建
Main Entrance Installation

FIRE
HOSE
REEL

南山婚姻登记中心 深圳

Nanshan Marriage Registration Center, Shenzhen

设计时间 / Design：2008 ~ 2011
建成时间 / Completion：2011
用地面积 / Site Area：3002.5m^2
建筑面积 / Building Area：977.5m^2
项目组 / Design Team：孟岩 | 朱加林 吴文一 | 傅卓恒 张震 | 魏志姣 |
王俊 胡志高 尹毓俊 李强 张新峰 | 廖志雄 林挺 于晓兰 刘洁
合作 / Collaborators：郭群设计（室内） 广州容柏生建筑结构设计（结构）
天宇机电设计（机电） 深装总装饰工程（施工图） 黄扬设计（标识）
业主 / Client：深圳市南山区建筑工务局 深圳市南山区建设局
建筑摄影 / Photographer：吴其伟 孟岩

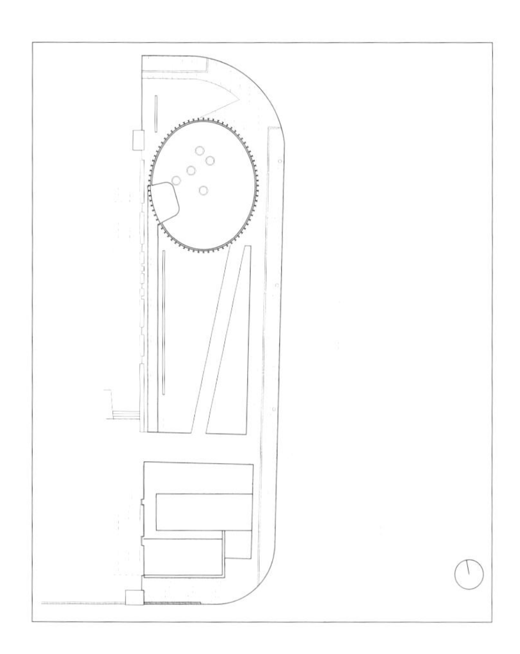

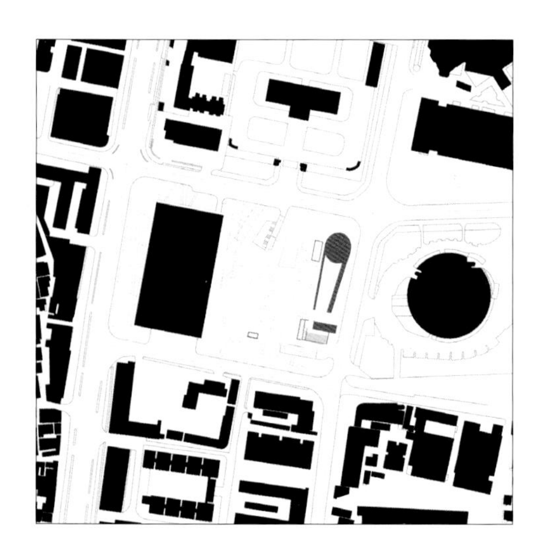

在中国的现实生活中，婚姻登记处作为民政部门的一个办事机构，只是个平常和平淡的场所，原本浪漫和令人激动的婚姻登记变得枯燥和程序化了。作为婚姻登记处——一个新的建筑类型，南山婚姻登记中心不仅能够为前来登记的新人们带来新的生活体验，更能成为一个信息发布的媒介，展示和记录新婚夫妇登记结婚的这一美好历程，同时，也为城市创造一个留存永久记忆的场所。

项目基地位于深圳市南山区荔景公园的东北角，长约 100 米，宽约 25 米。位于北端、靠近街道转角位置的建筑主体，通过架在水面上方的浮桥，与基地南端的凉亭广场相联系。这种布局方式不仅强调了结婚登记的仪式感，也使得位于街角的建筑主体成为一处具有象征性的城市标志物。

人们在建筑中的特殊体验是这个项目设计的重点。建筑内部的一条连续的螺旋环路舒缓地串联起整个序列性的片断：到达、在亲友的注目下穿过水池步向婚礼堂、合影、等候、办理、拾级、远眺、颁证、走过坡道、穿过水池、与等候的亲友相聚。在建筑的内部空间，以需要相对私密的小空间来划分完整的空间体量，在剩余的充满整个建筑具有流动性质的公共空间之间形成通高与镂空等丰富的空间效果。

包裹整个建筑主体的表皮由两层材料构成：外表皮的铝金属饰面用细腻的花格透出若隐若现的室内空间；内表皮则由透明玻璃幕墙构成真正的围护结构。整个建筑内部空间和外部表皮统一的白色烘托出婚姻登记的圣洁氛围。

In China, the marriage registration office's image is closely linked with the Government. In reality, the Registry is an office of the civil affairs department, so it is normally perceived as a common and dull place, as part of the bureaucracy. This situation turns the supposedly romantic and exciting idea of marriage registration into a routine and boring experience. Nanshan Marriage Registration Center is a new architectural type, for which the architects hope to bring new life experiences to new couples, and to create a medium for information display, recording of newly registered couples, and also retain for the city a permanent memory of the journey of marriage.The site of the project is in Lijing Park in Nanshan district, located in the Northeast corner of the park, approximately 100 meters long and 25 meters in width. The main building is placed in the northern side of the site, close to the street corner. A small pavilion on the southern side is connected with the main building by two bridges floating on a reflecting pool. The overall arrangement reveals this series of ceremonial spaces gradually. At the same time, it also makes the main building a symbolic civil landmark. A key point of this design is to discover how to organize the personal ceremonial experience. A continuous spiral shows part of the process in the whole sequence—"arriving, approaching to the wedding hall with the focus of relatives, photographing, waiting, registering, ascending, overlooking, issuing, descending slope, passing the water pool, and reuniting with relatives". For the design of the building, the whole volume is divided into smaller spaces to achieve relative privacy. The remainder of the whole building is full of a flow that creates a rich spatial effect. The building's skin is separated into a double layer structure, with the first layer using a floral mesh aluminum to reveal the interior, and the second layer using glass walls to provide a weatherproof structure. The overall inside space and the outside façade are all white in order to show the saintly atmosphere of marriage registration.

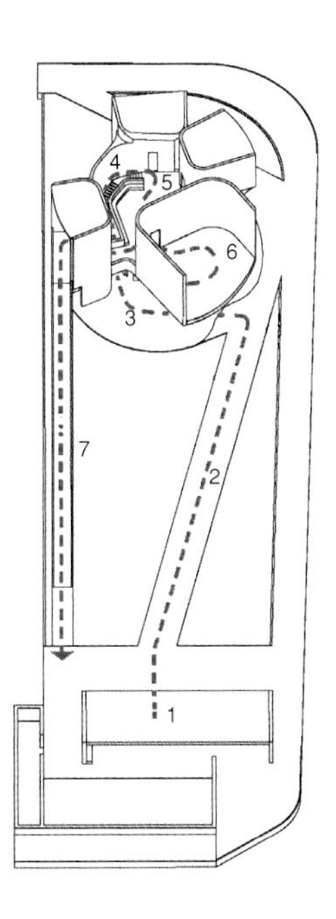

流线示意图
1 前亭 Pavilion
2 步道 Path Way
3 接待处 Reception
4 楼梯 Stairs
5 等候区 Waiting Area
6 颁证室 Marriage Certificate Room
7 坡道 Ramp

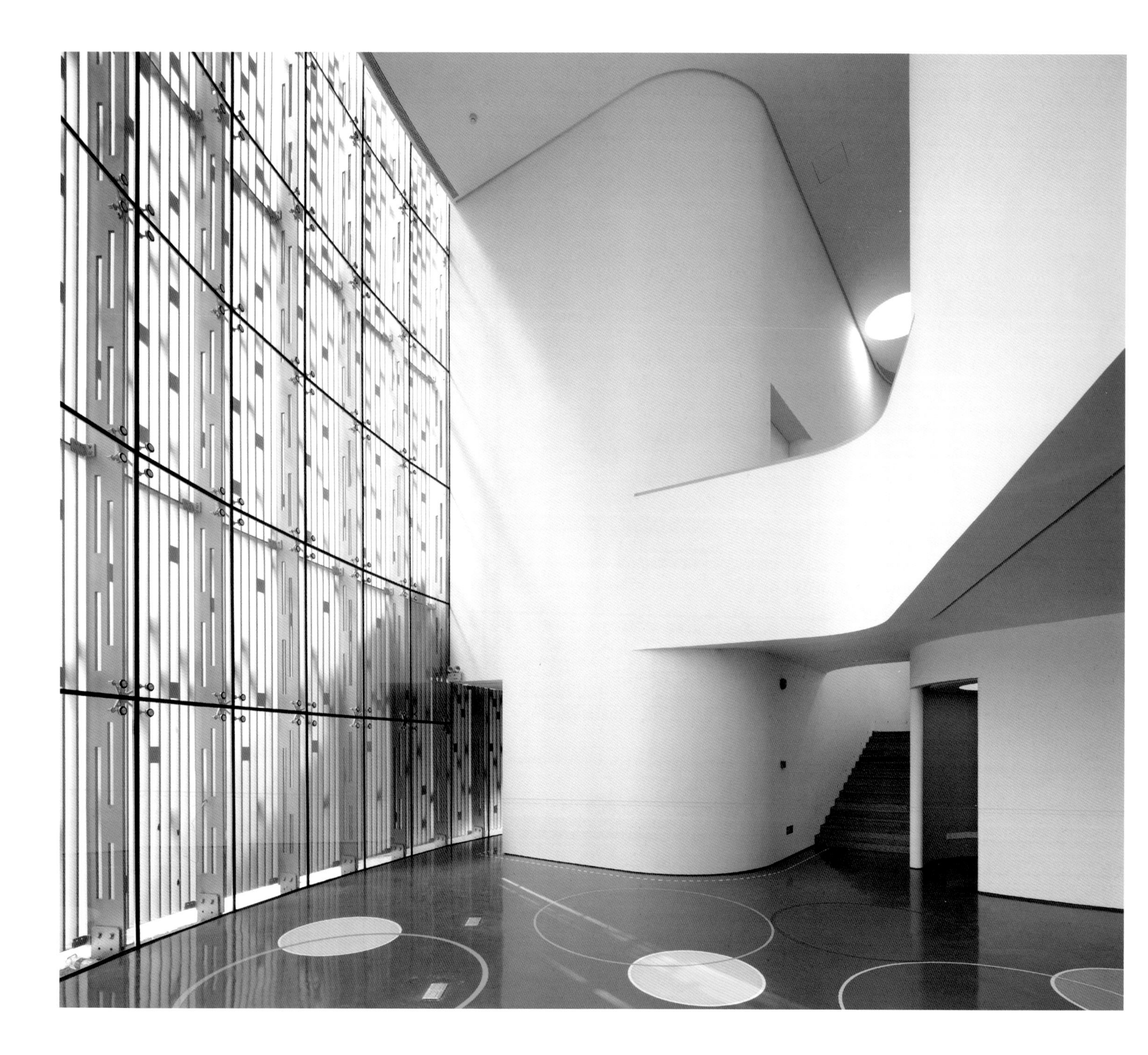

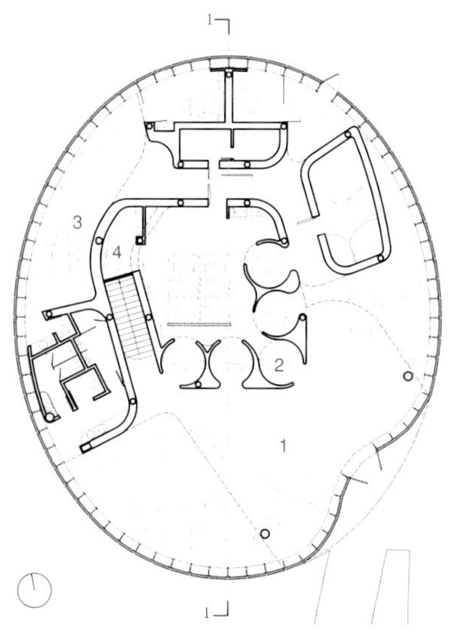

一层平面图 / The 1st Floor Plan

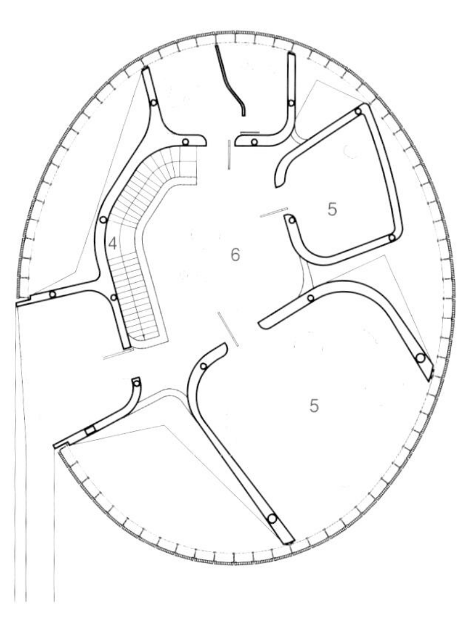

二层平面图 / The 2nd Floor Plan

1 接待 Reception
2 登记室 Registration Room
3 会议区 Meeting Room
4 楼梯 Stairs
5 颁证室 Marriage Certificate Room
6 等候区 Waiting Area

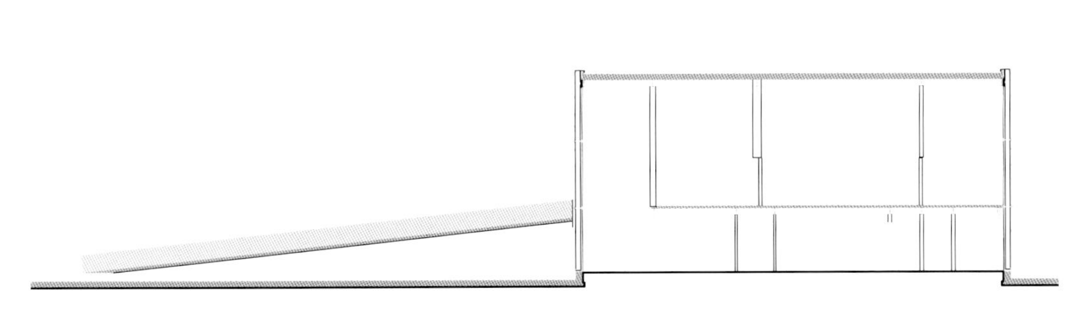

1-1 剖面图 / Section

回酒店 深圳

Hui Hotel, Shenzhen

设计时间 / Design：2010 ~ 2011
建成时间 / Completion：2011 至今
用地面积 / Site Area：1937m²
建筑面积 / Building Area：10915m²
项目组 / Design Team：刘晓都 | 姚殿斌 | 王俊 张震 | 李强
李嘉嘉 梁广发 姚晓微 林怡琳 姜轻舟
合作 / Collaborators：杨邦胜室内设计（室内）
同济人建筑设计（结构） 筑道建筑工程设计（机电）
业主 / Client：深圳市回品酒店有限公司
杨邦胜酒店设计顾问有限公司
摄影 / Photographer：吴其伟

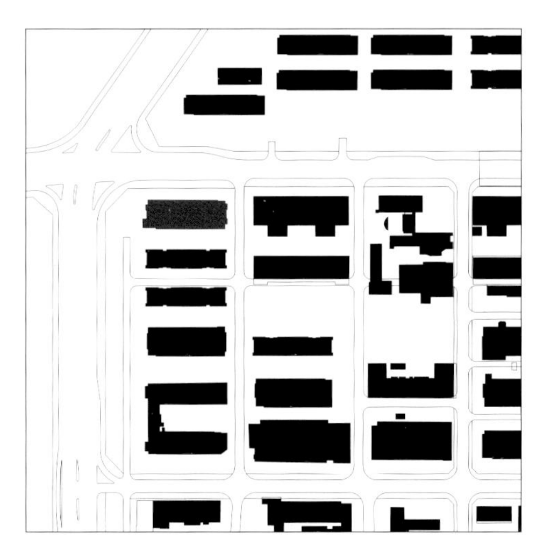

项目基地位于深圳市华强北片区的西北角，20 世纪 80 年代中期，这里曾是以生产电子、通信电器产品为主的工业区。深圳城市的飞速发展，使这个区域从工业区转型为繁华的超级电子交易商圈。基地上原有的工业厂房，因其单调陈旧的建筑形式，早已成为与该区域飞速发展不相符的残留物。规划部门希望保留此片区原有的肌理，对原有厂房采取改造而不是拆除的策略。

该项目投资方是家高端服装品牌设计公司，考虑到基地所处的优越地理位置具有极大的商业潜力，决定将其改造为酒店，希望建筑师能给酒店做出特殊定位，体现前卫、时尚、独特的企业精神。基地四面的景观资源差异甚大：北面临城市主干道，为最重要的迎街面，原有立面的窗墙体系已不能适应高端客房视线要求，其改造策略为采用视野开阔、大小不一的凸窗。以标准客房单元为造型元素，窗户向外偏心凸起并作导角处理，与墙面平滑相接，形成四种基本模数，再通过重复、旋转、镜像、变异等手法排列出动态变化的三维立体效果。西面处在街道转角，面向优美的城市中心公园，视野开阔，在此面处理上将凸窗变异，扩大客房的开窗面积，将最高级套房设置在此，让客人能够享受到最好的自然景观。而南面为酒店的背面，面对陈旧的厂房宿舍，无景观可言，同时存在严重的对视问题，因此采取了侧向开窗的造型单元，在满足自然采光的同时，避免客人在客房里正视对面极为杂乱的景象。

The project sits at the northwest corner of Huaqiangbei district in Shenzhen, which used to be a manufacturing basis for electronic and media communicating devices in the 1980s. The recent rapid development of Shenzhen's urbanization progress has transformed this particular area from a simple industrialized manufacturing basin into a vigorous digital commercial center. The original factory buildings remaining on site, with their disrepaired structures and generic monotonous building forms, have become left-over residues that do not accommodate the future development of the city.

The planners suggested preserving the original urban fabric of the site to incorporate a strategy based on revitalizing and regenerating the existing buildings, rather than a tabula rasa strategy. Being a high-end fashion design corporation, the client considered the site to have vast commercial potential with its excellent location; the project has been defined as a design themed hotel. The architect was asked to provide a distinct character which would elaborate the cutting-edge fashion expression with the corporate's creative-oriented culture.

The most important façade which defines the hotel image is on the north, facing the main city road. The façade opening modules are re-designed to adapt hotel room viewing purpose. Using the typical hotel room opening as designing module elements, the windows are projected away from the façade to provide maximum viewing possibilities, and the frames are off-centered then filleted smoothly back into the solid wall panels. This modular element is then repeated, rotated, mirrored and deformed to array a dynamic three-dimensional effect.

The same module on the west façade is repeated and enlarged; the ensuites units are planned on the west end to provide the best viewing direction towards the city park. Contrarily, the south façade, facing a factory dormitory with no viewing pleasure at all, was ensured with privacy and natural lighting, in order to facilitate the hotel program.

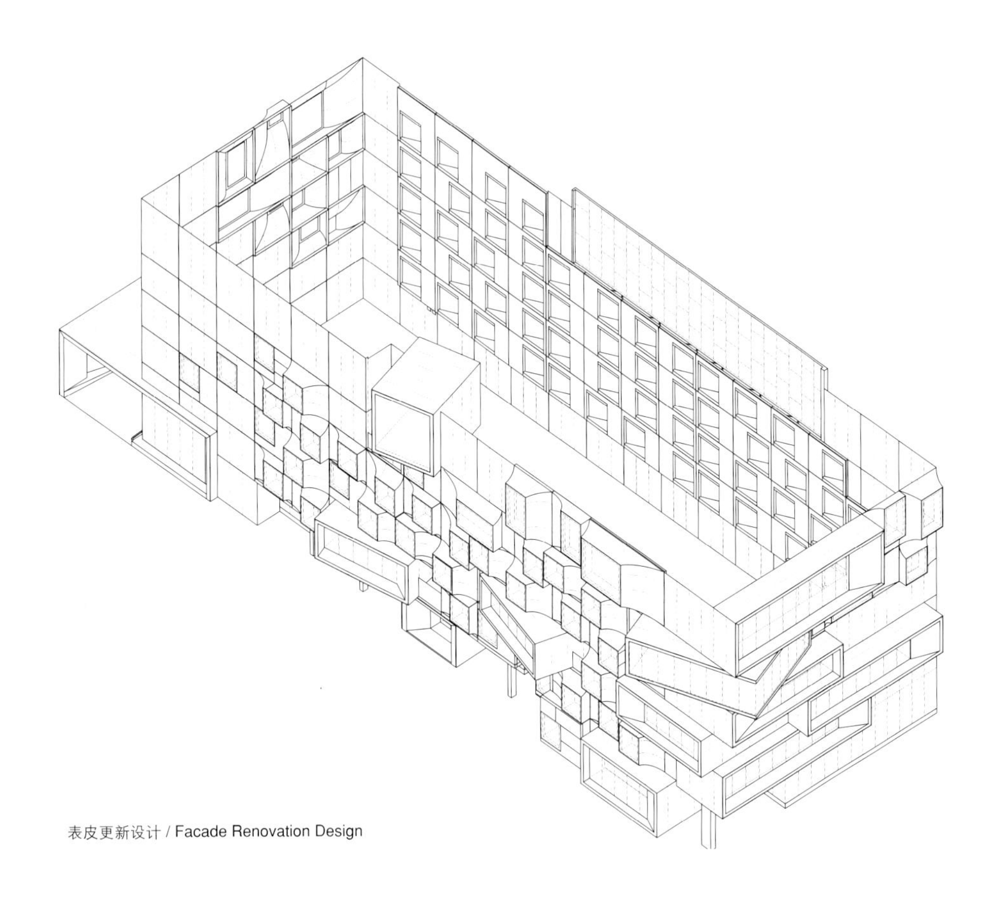

表皮更新设计 / Facade Renovation Design

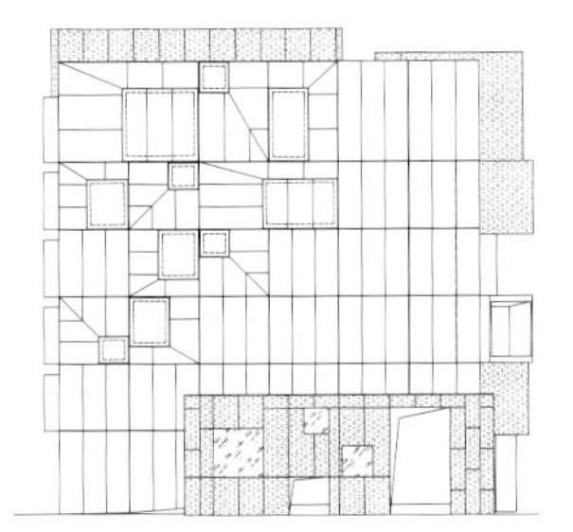
东立面图 / East Elevation

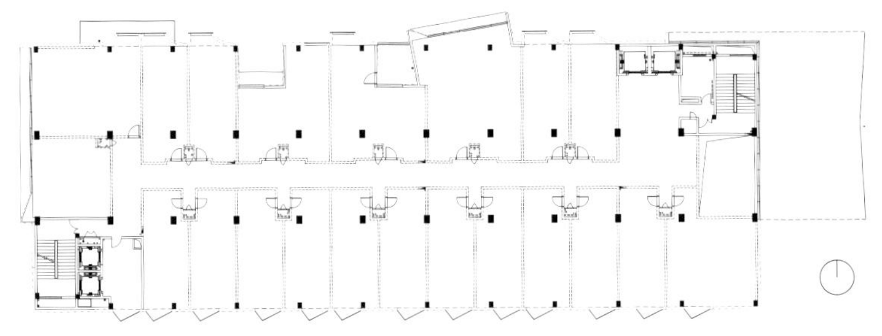
二层平面图 / The 2nd Floor Plan

白云观珍宝花园 北京

White Cloud Temple Jewelry Garden, Beijing

设计时间 / Design：2006 ~ 2008
建成时间 / Completion：2008
用地面积 / Site Area：$13660m^2$
建筑面积 / Building Area：$12684m^2$
项目组 / Design Team：王辉 刘旭 | 郝钢 王鹏 苏立恒 |
魏燕 陈春 刘爽 Kristin Klopfenstein
合作 / Collaborators：华航建筑设计公司（施工图）
业主 / Client：华航房地产开发建筑公司
摄影 / Photographer：杨超英 陈尧

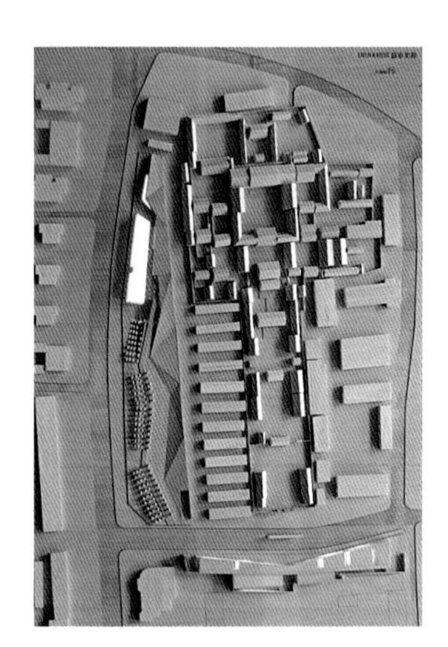

白云观在道教历史中具有举足轻重的地位，现今是中国道教协会的总部所在地。随着城市的拆迁，包裹在密集民宅中的古老道观被剥离出来，失去了原先应有的环境依托，被迫与当代城市对话。

本项目位于沿道观西侧和南侧的狭窄地块，定位为高端商业。然而西侧用地的三分之二已被规定为城市公共绿地，两块地之间的建筑很难呼应。面对种种挑战，本设计立足于将当前的文化意识和古老的文明传统对接，而不是重复历史的语言来获得新旧建筑之间简单的和谐。设计目标是使新建筑成为白云观嵌入当代城市的镜框。

对应于白云观自身的中轴线，是一道微折成"V"字形的、约140米长的现代玻璃影壁墙，它折射了道观的重重庭院。其立面构图是道教八角卦形图案的变形，由此衍生出可以组合成门、窗、橱窗、灯箱、不透明墙体等的复杂幕墙体系，既有功能的合理性和视觉的美感，又有文化的厚度，从而以轻松又有力度的现代构图应对古观的凝重。平行于白云观主轴线的西侧，是一条曲折的绿轴，它把规划中的绿地延伸到商业的屋顶，从而使这部分的建筑弱化，形成了与白云观庄重的院落节奏感相呼应的自由和弦。这一和弦的伴奏，从精神上契合了故宫和三海之间那种亦庄亦谐的中国城市规划的风韵。由于建筑高度的限制，建筑的主体部分在地下，设计中充分使用烂漫的空间和光影，把人流引入到活跃且有文化感的地下商业空间。

这个设计证明，在新的都市建设背景下，可以用当代意识来延续历史文化。

The White Cloud Temple has played an important role in Taoist history, and today it serves as the headquarters of the Chinese Taoist Association. Following the changes within the city, the ancient Taoist temple, previously wrapped in a dense traditional residential area, has lost its original environment and is forced to dialogue with a contemporary city.

This high-end commercial project is located along the narrow strips on the west and south sides of the temple. On the west strip, two-thirds of the site is reserved for public green space, thus disconnecting two commercial parts. In the face of all challenges, the design intends to bridge the old culture and the modern one with a contemporary architectural language, and to embed the existing temple into the contemporary city with a new frame.Corresponding to the White Cloud Temple's central axis, a subtle V-shaped glass wall approximately 140m in length is designed to reflect the layers of the temple's courtyards. The pattern of the façade is derived from a reorganization of a Taoist octagonal theme, and the pattern fits the functionality of doors, windows, light boxes, etc., with a visual delight as well as a cultural flavour. Parallel to the main axis of the White Cloud Temple on the west side, there is a zig-zag green axis that extends the garden to the roof of the commercial area, so as to soften the presence of the new building's massive volume. This park makes a harmonic melody into the symphony of the temple's main chord progression. This melodic accompaniment gives the charm of the relationship between the formal and informal, which is similar to the relationship between the Forbidden City and the bodies of water to its west side. Due to height limitations, a large volume of the program is put below grade. The design makes great effort to draw the public into the underground space through a romantic language of light and space.This design asserts that in line with the new urban development, history should be rejuvenated by a new cultural consciousness.

一层平面 / The 1st Floor Plan

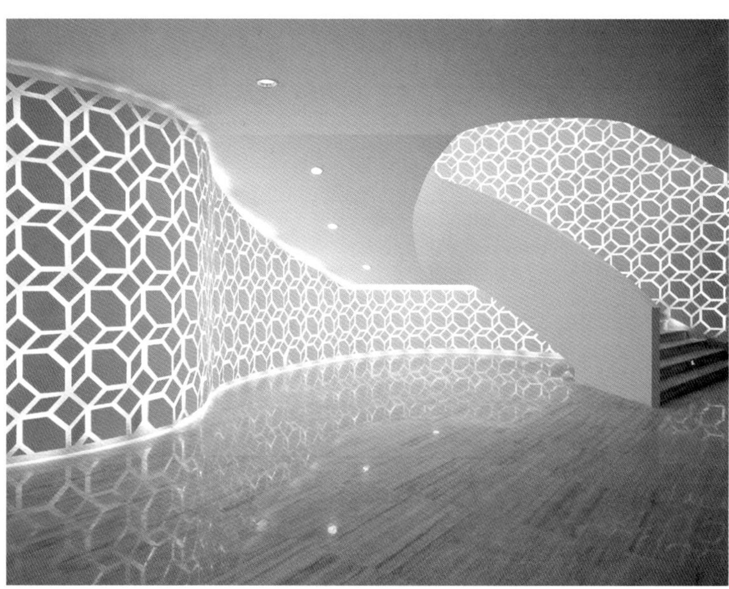

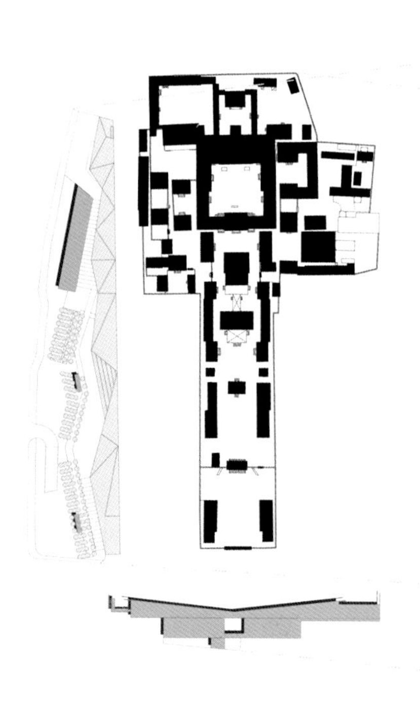

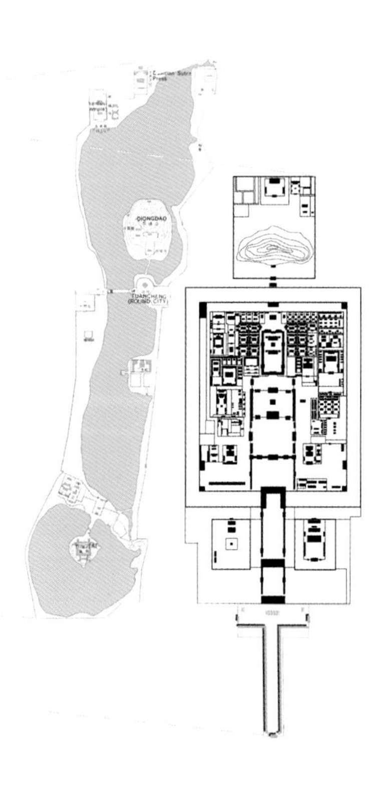

概念示意图 / Concept Diagram

唐山城市展览馆及公园

Tangshan Urban Planning Museum and Park

设计时间 / Design：2005 ~ 2006
建成时间 / Completion：2008
用地面积 / Site Area：22850m²
建筑面积 / Building Area：5900m²
项目组 / Design Team：王辉 刘旭 | 张柳娟 陈春 | 刘爽 王鹏 段云龙 赵利卫 王玮
合作 / Collaborators：北京皇都建筑设计（施工图） CBDC 中房钢构（钢结构）
业主 / Client：唐山市规划局
摄影 / Photographer：杨超英 孟岩

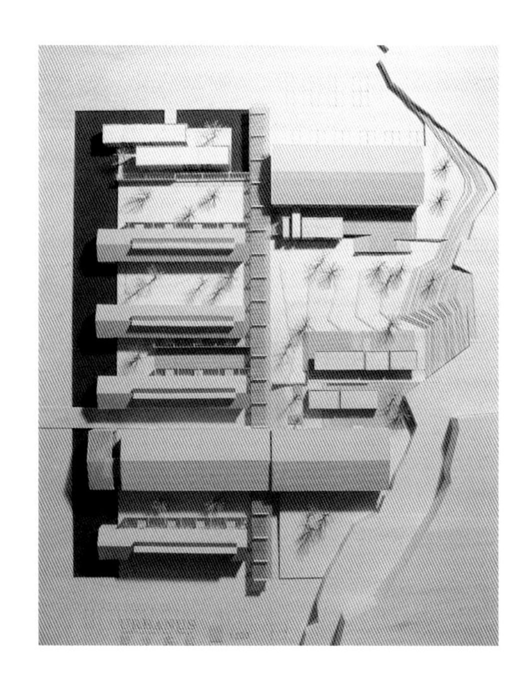

唐山城市展览馆及公园的前身是位于大城山西麓的原唐山面粉厂。工厂在新的都市发展条件下搬迁后，用保留下来的四栋日伪时期的仓库和两栋地震后建的仓库作为展示厅，改造成一个留有城市记忆的博物馆公园。

保留下的六栋平行的建筑恰巧垂直于山体，构成了一种有意味的韵律。改造设计通过非常少的加建，更强化了这种韵律，使山有节奏地从建筑间的空隙中溢到城市，形成了"大城山—山脚后花园—厂房间小院—大公园—城市主干道"一系列有层次和有序的城市开放空间体系。加建部分的笔墨不多，通过新旧的材料对比，对老仓库的屋顶和门廊夸张重构，用水池和连廊来统一离散的个体建筑等处理手法，精心地呵护和放大厂房和山体间构图上的天作之美，在新的环境下彰显毫无美学价值的原有建筑群的内在美。

对于唐山这种面貌庸常的城市而言，当前的城市化运动和当年的地震一样，都在使唐山平凡的现当代建筑史消失。但对于这个震后重建的城市，平凡是一种朴实的城市精神和恬静的城市生活方式，是一种一般市民可承受的体面环境。它不逊于一线城市火热的生活。当人们走进展览馆去畅想城市未来后，走出展馆又能体会城市的历史。这种平衡有助于去思索今天的城市化行动所应有的策略、立场和价值观。这个建筑提醒着城市建设的执行者，城市更新不是地震式的颠覆。

Tangshan Urban Planning Museum is located in the old flour mill on the western slope of Dacheng Hill in the downtown area. After the relocation of the factory, the area was transformed into a park. The original four warehouses built during the Second World War and another two built after the earthquake, are preserved as showrooms in a museum park keeping the memory of the old city.

The six parallel buildings are rhythmically perpendicular to the hill. The new additions are designed to intensify and organize this rhythm, as well as to link the hill to the city with a rhythmic pattern: the hill – the garden at the foot of the hill – the courtyards between the warehouses – the museum park – the urban street. Through using new materials in contrast with the old, exaggerating the old warehouse profile, connecting individual structures with the pool and corridor, and not disturbing the visual relationship between the warehouses and hill, the additions made to the existing site are minimal so as to highlight the beauty of the original structures.

In recent years, the urbanization movement has been as devastating as the 1976 earthquake in Tangshan, as it erases the city's contemporary architectural history. Unlike life in the fast-paced first-tier cities, the rebuilt city has a spiritual simplicity that responds to the simple local lifestyle. The essence of architectural preservation is to preserve such simplicity. The planning museum fulfills this task. Inside the exhibition hall, visitors can imagine the future of the city, and outside of the pavilions they can still sense the city's history. This balance is helpful to evaluate the strategy, stance, and value of new urban construction strategies. This design asserts that urban renewal is not an unstoppable earthquake.

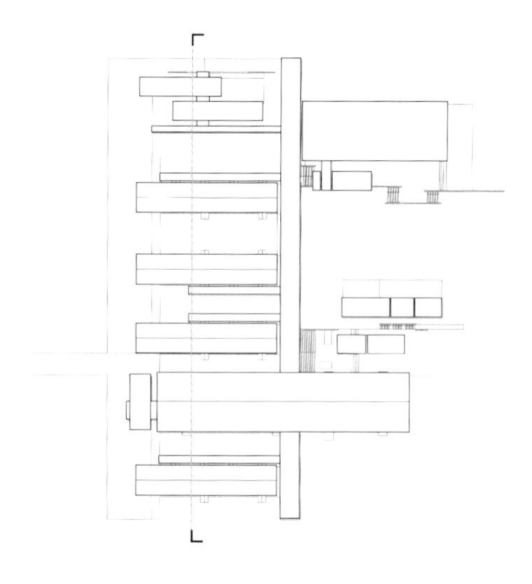

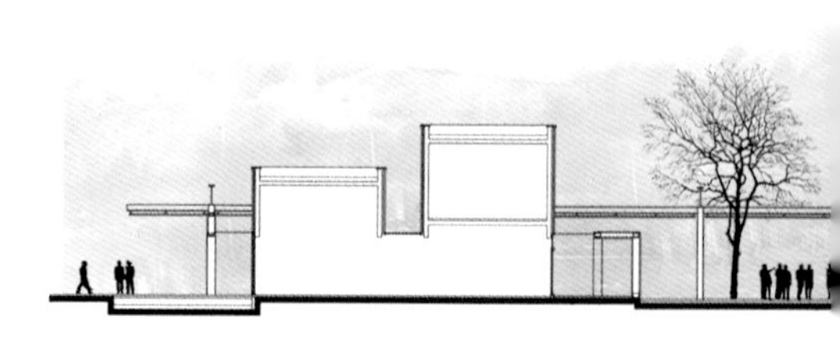

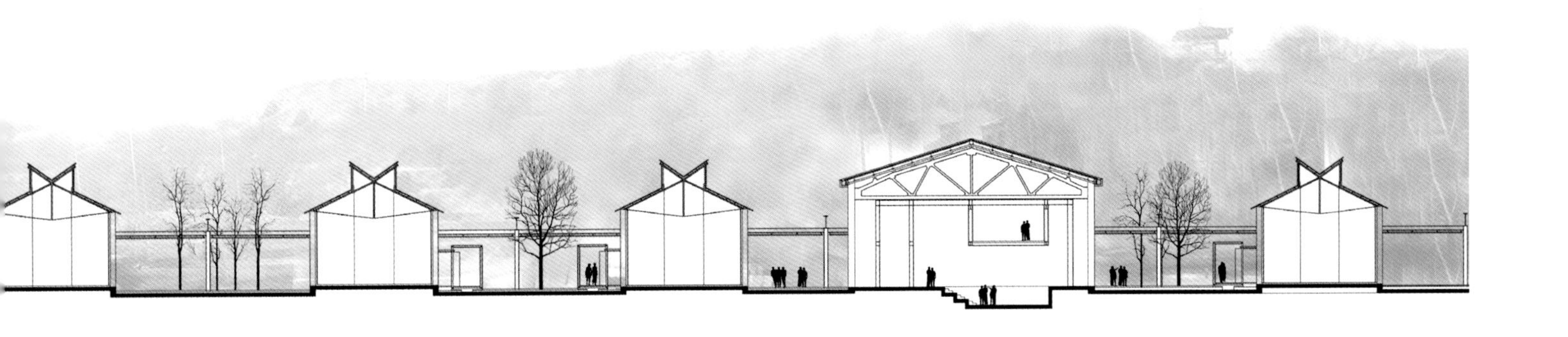

D

B

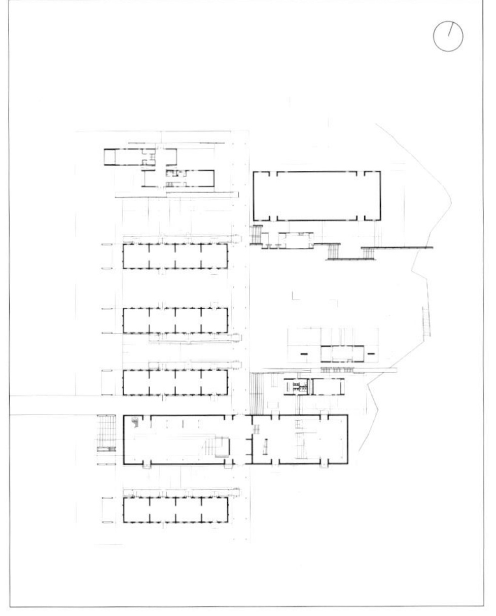

一层平面图 / The 1st Floor Plan

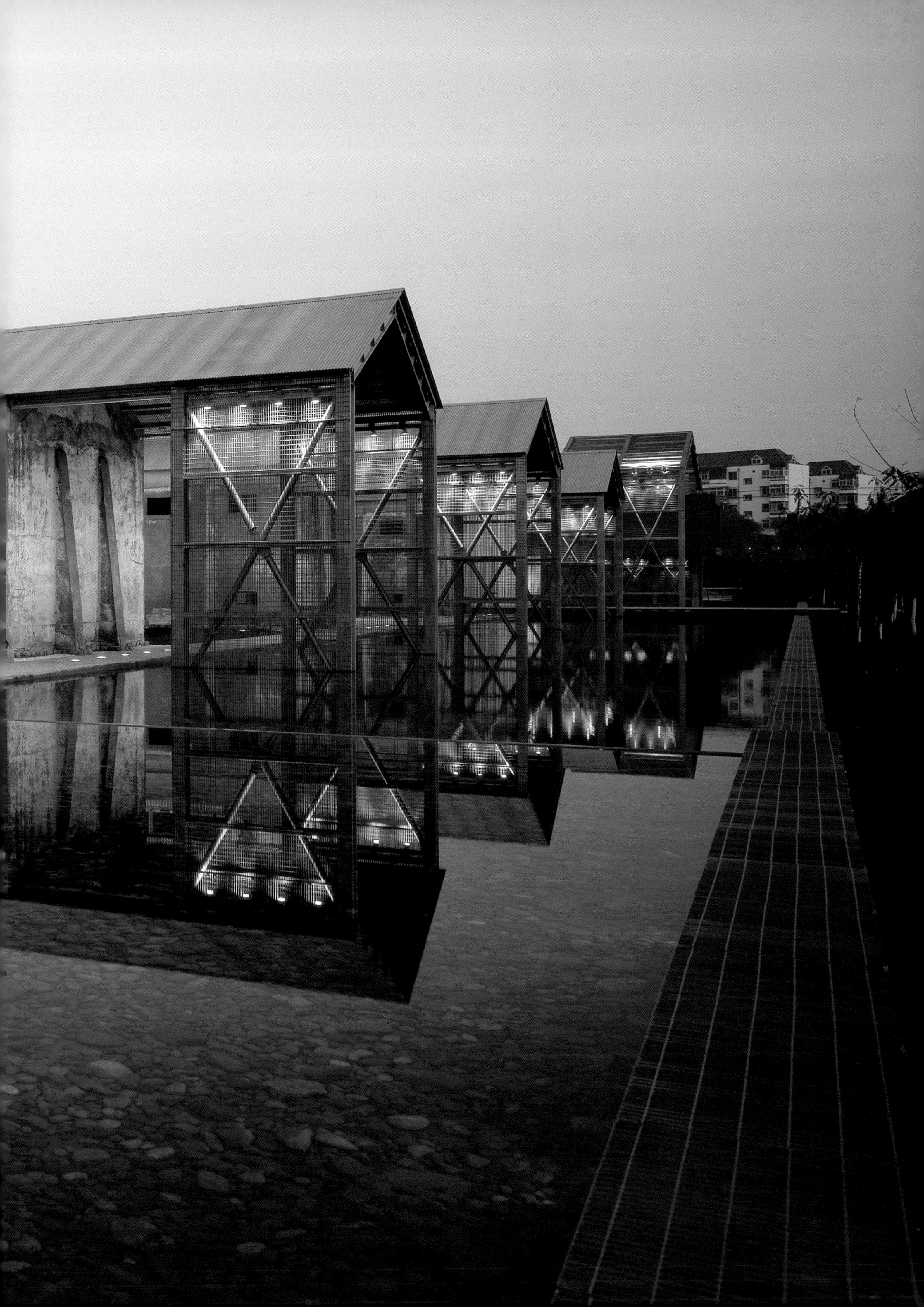

唐山博物馆改造扩建

Tangshan Museum Expansion

设计时间 / Design：2009
建成时间 / Completion：2011
用地面积 / Site Area：20000m²
建筑面积 / Building Area：24444m²
项目组 / Design Team：王辉 | 吴文一 刘银燕 | 杜爱宏 郝钢 |
张永建 张淼 成直 郑娜 陈春 魏燕 刘爽 刘妮妮 杨勍 陈岚 霍振舟
合作 / Collaborators：北京龙安华城建筑设计（施工图） 无界景观（景观设计）
业主 / Client：唐山市政府投资工程建设管理中心
摄影 / Photographer：陈尧 郝钢

新唐山博物馆的前身是建于1968年的毛泽东思想胜利万岁展览馆。这个呈"品"字形的建筑群也是1976年唐山大地震后幸存的为数不多的大型公共建筑，因此旧建筑本身成了新博物馆的最大收藏品。该建筑群位于市中心的凤凰山脚下，其开阔的广场也是城市中最为市民喜爱的公共空间。

改造后的新博物馆，通过有机的动线组织，把三座离散的建筑组织成一体；通过完善的功能配置，实现了一个现代化的博物馆；通过在漫长的展线中穿插进许多非展陈性的开放厅堂，使这个免票的博物馆成为供市民休闲和交往的客厅。新加建的部分弥补了旧建筑体量上的比例失调，同时依然保持旧建筑作为这个建筑群的主体。采用超白彩釉玻璃这一现代材料，既拉大了新旧建筑之间的时代距离，又弱化了加建部分的体量，从而突出了旧建筑。新建部分放在边角位置，没有破坏旧建筑和背景山体之间的既有视觉关系，从而保持了市民对这个城市场所的记忆。广场的改造使用"城市沙发"的概念，在开敞的公共广场中植入半私密的、适合小群体活动的个人化空间，使人们在这里有领域感和归属感。

这个设计探讨了如何有机地保持当前二、三线城市中能代表城市记忆的建筑物，让这个锚固在自己的精神原点上的城市在新的社会文化、经济和技术条件下，实现一个美丽的转身。

The Tangshan Museum is a renovation project based on the original 1968 pavillions for learning derived from Chairman Mao Zedong's ideology. The 1968 structures are arranged in a "C"-shape around a large public square. These buildings have survived the 1976 Tangshan earthquake, and have become the largest collection for the new Tangshan museum. At the foot of Phoenix Hill, the site sits at the heart of the city and its open square is the city's most favorable leisure and recreational destination.

The new Tangshan Museum intends to integrate the three existing buildings into a wonderful exhibition venue. The museum is modernized with a thoroughly functional program, and it is easily accessible through a series of halls free to the public. In terms of volume, the new addition modifies the old buildings' proportional relationship, while maintaining the old parts as the main visual focus. By using modern light materials such as white ceramic fitted glass, the differences between the old and the new are accentuated.

The new part is located in the corner area so as not to disturb the existing visual relationship between the old buildings and the hill in the backdrop. This sensitive measure keeps people's memory of this place intact.Driven by the concept of "the urban sofa", the central square is designed with a ring of semi-private spaces suitable for small group activities, and thus creates a sense of belonging in the city.

The design explores how to maintain memorable urban artifacts in a second or third tier city. Under new social, cultural, economic and technological conditions, it seeks a strategy of how to transform the old into the new with the premise of anchoring the city into its spiritual roots.

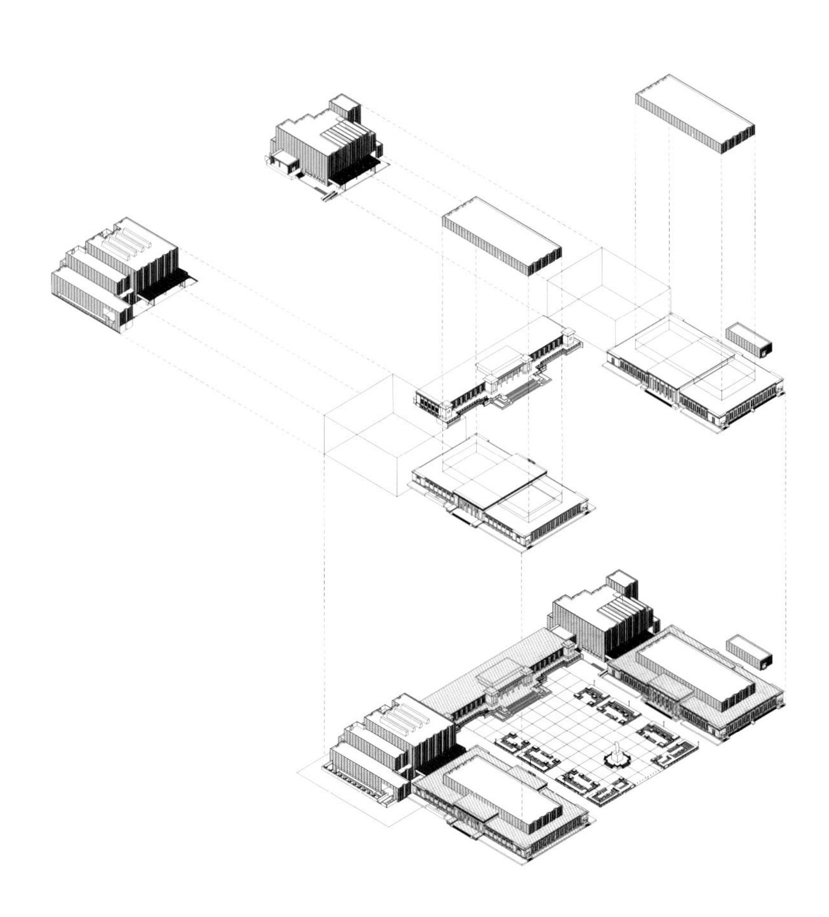

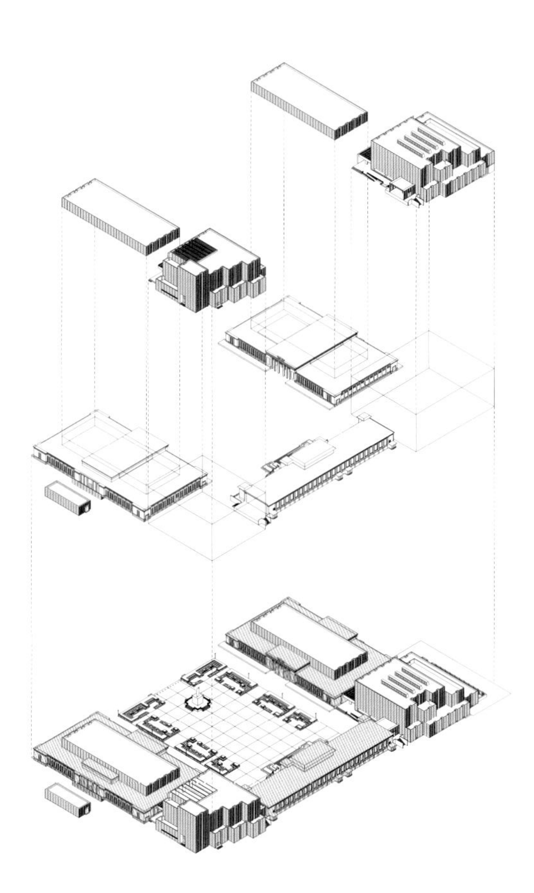

共创文明城市 建设美好家园
唐山市

活动

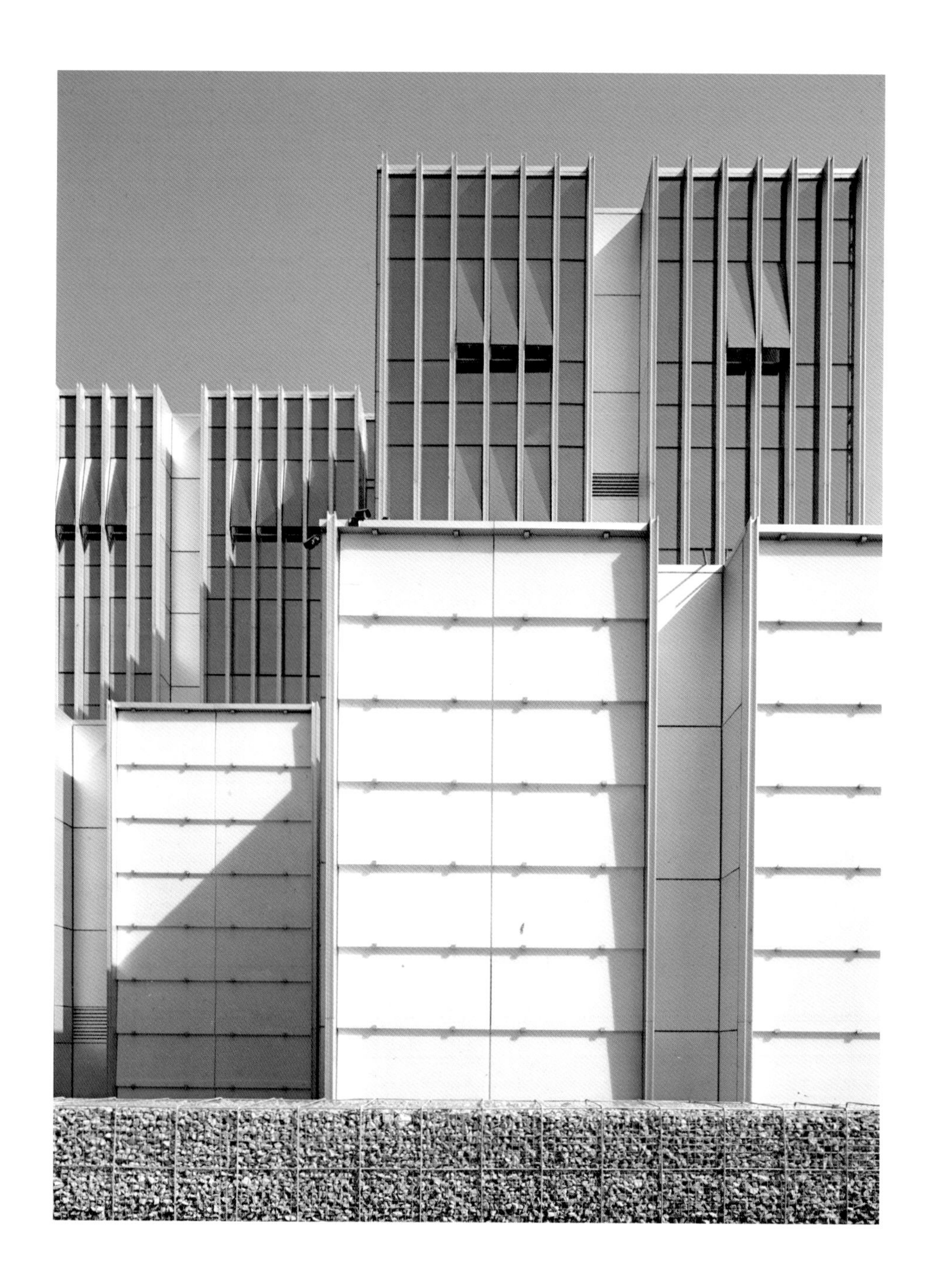

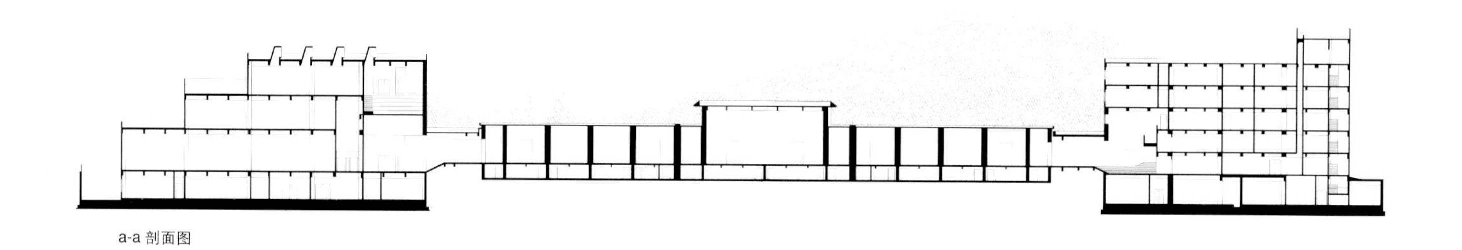

a-a 剖面图

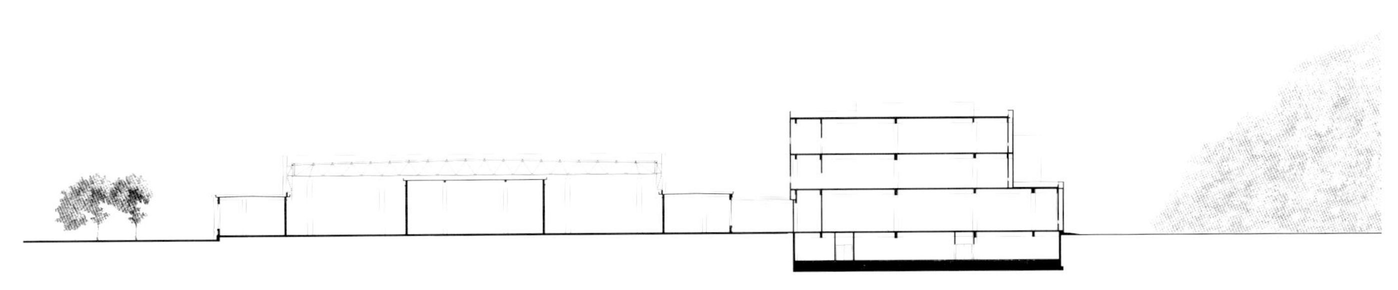

b-b 剖面图

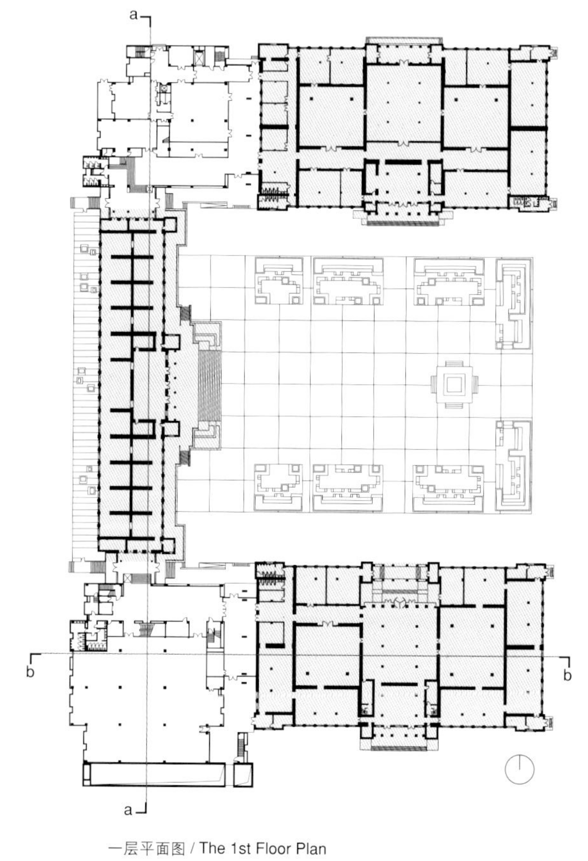

一层平面图 / The 1st Floor Plan

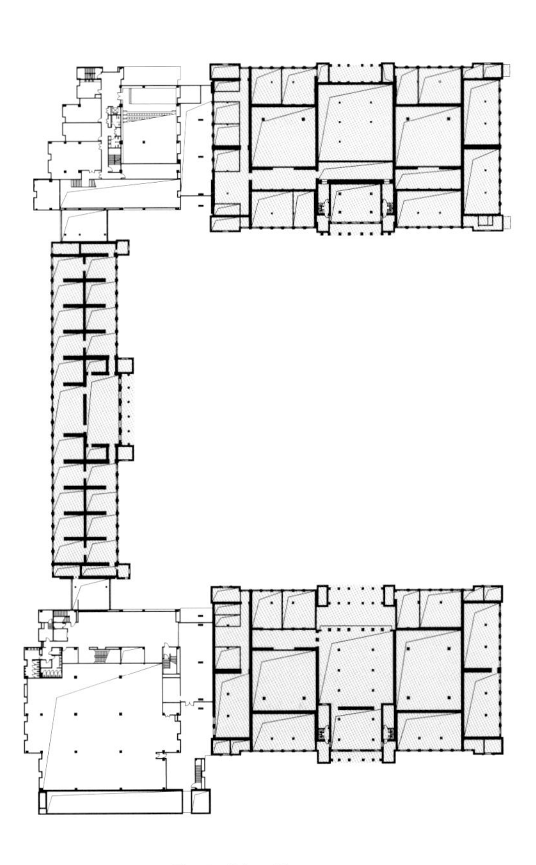

二层平面图 / The 2nd Floor Plan

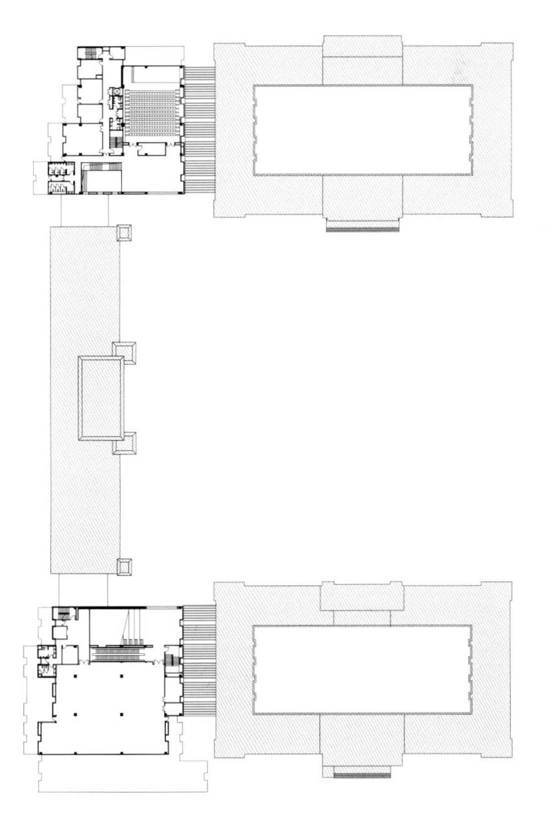

三层平面图 / The 3rd Floor Plan

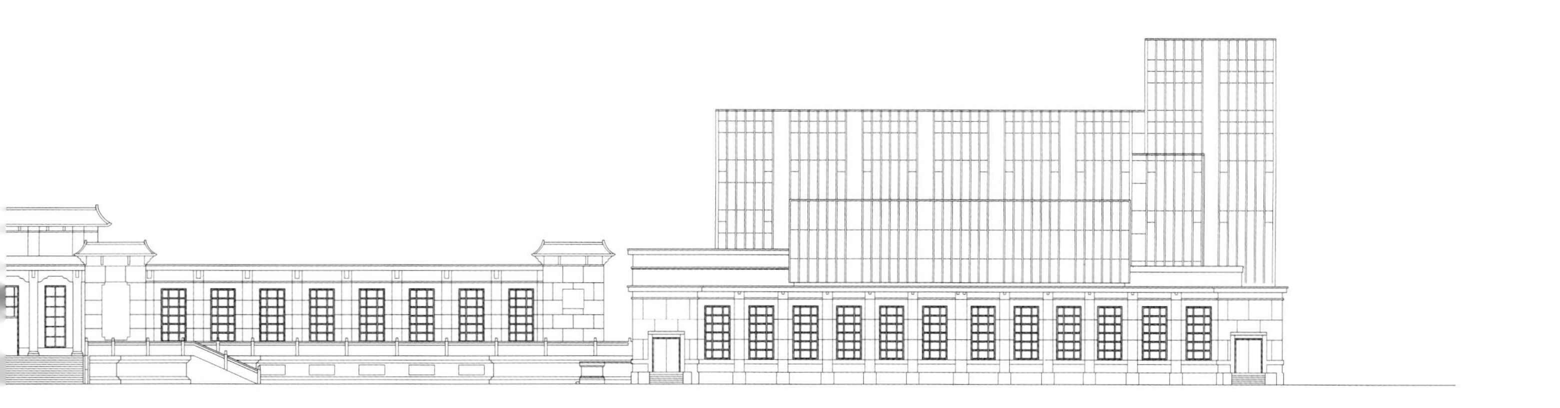

华侨城创意文化园北区改造 深圳

OCT-Loft North Area Renovation, Shenzhen

设计时间 / Design：2008 ~ 2011
建成时间 / Completion：2011
用地面积 / Site Area：32276m²
项目组 / Design Team：刘晓都 孟岩 | 傅卓恒 王衍 黄艺宏 魏志姣
吴文一 | 廖志雄 欧阳祎 熊嘉伟 丁钰 黎靖 何慧珊
合作 / Collaborators：易逊建筑设计（施工图）
深圳市建筑设计研究总院（钢结构）
业主 / Client：深圳华侨城房地产有限公司
摄影 / Photographer：吴其伟 孟岩 陈文赟

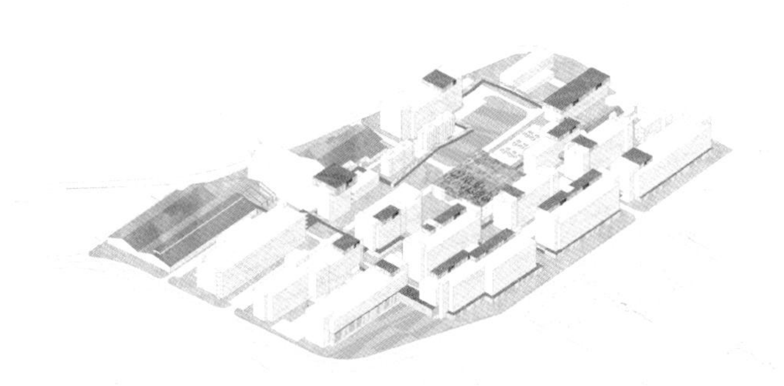

基地现存数十座建于 20 世纪 80 年代早期的厂房、仓库和宿舍楼。城市的高速发展，使这个位于中产阶级住区和迪斯尼般娱乐区夹缝中的旧厂区，渐渐变成不为人留意的都市残留物。2003 年，何香凝美术馆决定将其中一个厂房改造为当代艺术中心，这一艺术介入成为整个改造的启动点。

改造规划采用了置换与填充的思路，从在现有厂房中加入新艺术中心开始，整理厂区内可利用的结构，一步步添加和改造，融入以创意产业为主体的当代内容，使厂房被画廊、书店、咖啡厅、酒吧、工作室和设计商店渐渐填满。这些内容延伸、包裹、渗入到现有的肌理，创造了一系列相互贯通的公共空间和设施。通过这种拒绝一次性设计和开发的模式，让时间积淀出社区的厚度和底蕴。都市实践的总体控制与具体使用者个性发挥的互动十分符合创意产业的特点。

成功的创意园最大的潜能就是能促进不同类型的创意机构在各个层面进行交流。从这个角度出发，创意园第二期的设计升级了第一期南区的现有模式，在整体上控制宏观形象，进行业态组合规划，合理分区，控制比例，以设计学院作为园区可持续发展的源动力，将公共功能进行混合和叠加，利用连廊系统将学院的部分公共教室分布在各个楼中，将其设计成真正的创意交流场所，使得园区内部的交流行之有效和充满机会。这个改造实现了创意文化园的根本目标：将各方知识精英结合起来，鼓励跨领域、跨行业的对话和思想碰撞，开拓各种创意发生的可能性，使艺术社区成为凝聚艺术和设计创造力的基地。

The site currently has over dozens of early factories, dormitories and storages left vacant from the 1980s. The factory has a modest appearance in the midst of a mixture of middle- class residential areas and Disneyland-type entertainment zone. As time passed by, they became vacant and industrial activities declined.

In 2003, the Hexiangning Art Museum decided to set up a non-profit contemporary art center in one of these warehouses, called OCAT. The plugging-in of the OCAT set up an interesting paradigm in the beginning of the industrial area's regeneration. The master planning team's intention was to replace and fill up the buildings by applying the new spatial form with small-scale operations and improvements on the infrastructure. They started by adding programs to existing structures to adapt to the function of the art centre; the empty lots between them are intended to be filled with galleries, bookshops, cafes, bars, artist ateliers and design shops, along with lofts and dormitories. These new additions fill the open spaces and set up new relationships between buildings by wrapping and penetrating the existing urban fabric. They also create a second layer of urban spaces which tries to set up a dynamic, interactive and flexible framework to constantly adapt itself to the new conditions posed by the vast changes of the city.

At the beginning of the regeneration, the design team addressed the value of these industrial imprints and their extraordinary architectural quality, which provides flexibility for the coexistence of both industrial and creative occupants. The second stage of the renovation aims to reposition the overall branding image, the strategic planning is distinct from the south. The existing network, landscape and post-industrial buildings with the social network are interwoven to encourage public use. They enhance reappropriation of abandoned spaces, replacing them with new activities. The project intended to articulate and visualize the dialogue not only for the revitalization of the industry imprint, but also as a cohesion between new participants.

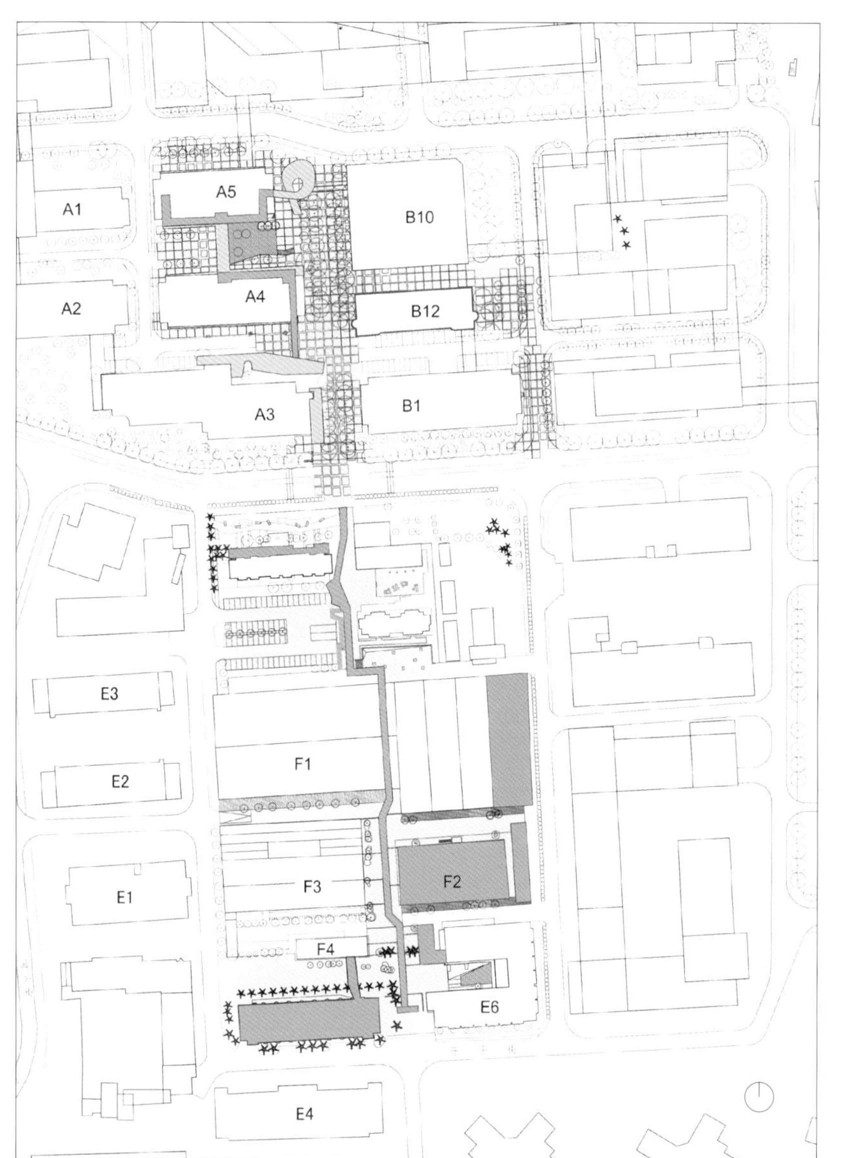
A1
A5
B10
A2
A4
B12
A3
B1
E3
F1
E2
F3
F2
E1
F4
E6
E4

A3

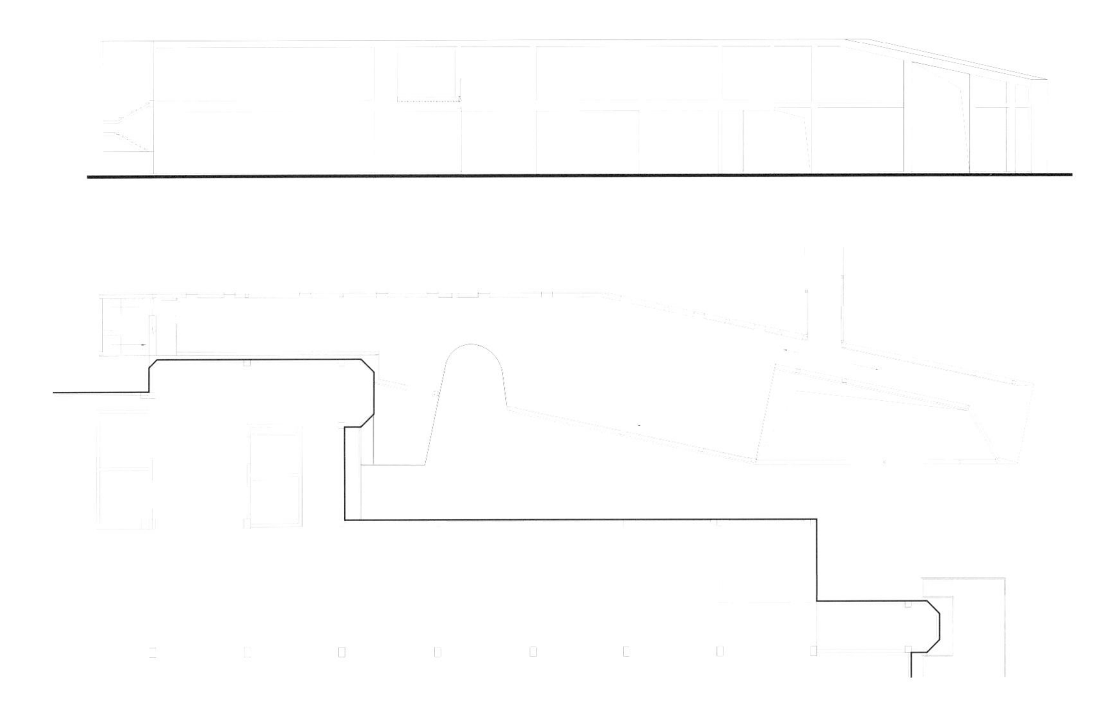

公共平台平面 / Public Platform Plan

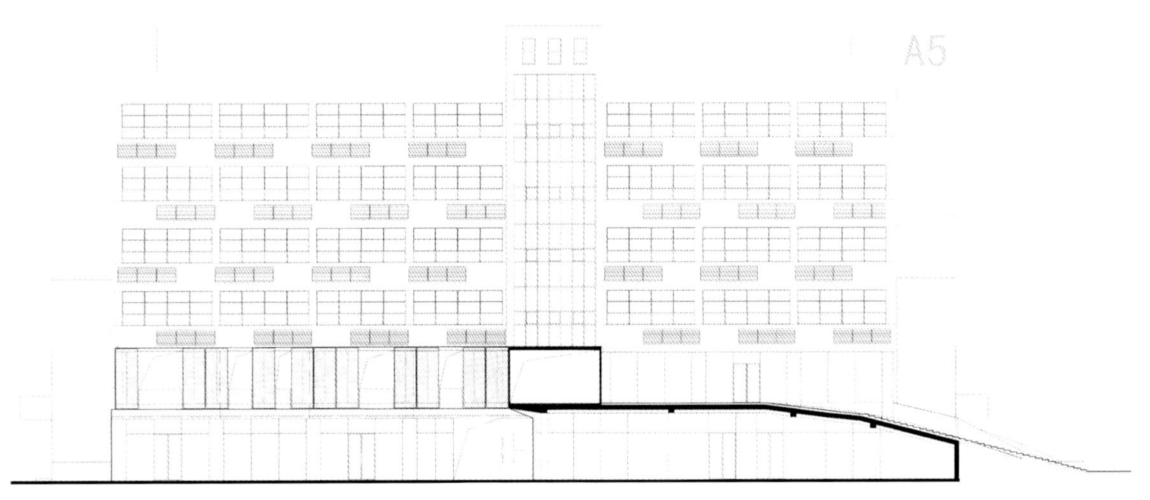

公共平台剖面图 / Public Platform Section

Contemporary Dance 2008
(2008)
OCT LOFT
创意市集

翠竹公园文化广场 深圳

Jade Bamboo Cultural Plaza, Shenzhen

设计时间 / Design：2005 ~ 2006

建成时间 / Completion：2009

用地面积 / Site Area：6871m²

建筑面积 / Building Area：地下车库 1433m²

项目组 / Design Team：孟岩 | 朱加林 | 邢果 廖志雄 | 刘小强 吴凯茂 黎靖 丁钰 魏志姣 左雷 夏淼 刘浏

合作 / Collaborators：北方 - 汉沙杨建筑工程设计（施工图）

业主 / Client：深圳市罗湖区翠竹街道办事处

摄影 / Photographer：孟岩

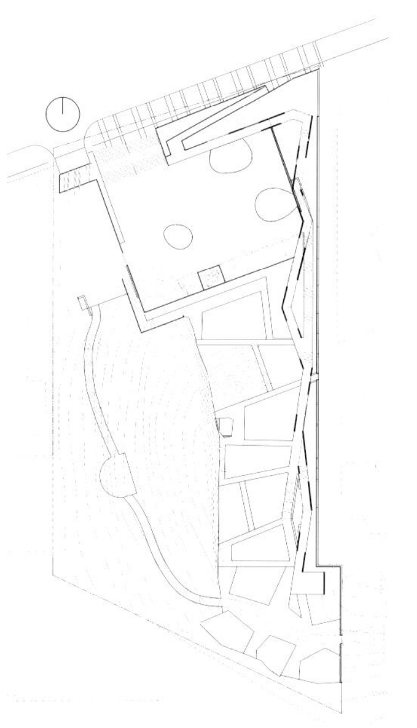

深圳在大规模开展城市建设以前，是丘陵地带。如今原始的地貌已很难在平整的城市格局中看到。翠竹公园与周边街道缺乏联系，是城市中心区里为数不多的、仍部分保留原始地形及植被的地段。

翠竹公园北入口借用临近的某高尚住宅区的一片开发残余剩地实现与北侧道路的连接。基地形状不规则，由北至南坡度高差有 13 米。设计保留了基地残留的小山丘地貌，采用传统中式园林的庭院形式，遍布的竹岛群为孩子们提供了捉迷藏、做游戏的地方，也为老人们围合出了下棋、打太极以及社区歌舞的场所。从庭院东北角出发，一条折线形长廊依现存的挡土墙而建，蜿蜒于山边，通向山顶，延伸到公园新的入口。折线形廊子与墙之间形成一系列三角形空间，重新界定了公园东侧的边界。竹、花、树，在这些被界定的空间里形成了若干幅中国画。行走于廊中，步移景异，这种系列的空间体验正是中国传统园林精髓所在。沿山体逐级抬升的长廊，将狭长的坡地切割成形状各异的种植台地。台地上栽种花、草、农作物，在鼓励附近居民和孩子们来参与体验种植的乐趣之时，也最大程度地引导公众参与社区环境的创建与维护。

从繁闹的城市生活跳跃到田园实践和竹林间的休闲活动，是对人们回归山野、远离尘嚣的内心渴望的积极回应。这种精神的回归之旅建立了一条将传统自然与现代城市链接的纽带。翠竹文化广场是带给人宁静致远的空间体验的现代中国园林，同时也是对深圳曾经的原始自然生态的一种见证和纪念。

Before large-scale urban construction began, Shenzhen was formerly hilly country. The Jade Bamboo Garden, which has insufficient connections with the surrounding streets, is one of the few areas partly conforming to the original landform and vegetation in the urban center. The north entrance of the garden is required to connect the roads towards the high-class residential development area.The site has an irregular shape, and the altitude difference of the slope from north to south is almost 13 meters. The design keeps the former hilly landform of the site, using the traditional form of the Chinese courtyards to create bamboo islands throughout. Such space allows not only for children to play hide-and-seek and other games, but also provides community services for the elderly such as chess club, Tai Chi practices and music performance. Starting from the northeast corner of the courtyard, an outdoor corridor shaped like a broken line is built along the existing retaining wall, climbing sinuously on the hillsides, leading to the peak, and extending to the new entrance of the park. Series of triangular spaces have redefined the east boundary of the park. Bamboo, flowers and trees are mixed together to form sceneries reflecting astonishing Chinese paintings. The scenes shift swiftly when walking, which accentuates the essence of the traditional Chinese garden.

The uplifted corridor cuts the narrow slope into several cultivated terraces with various shapes. Flowers, grass, and crops planted on the terraces encouraged local residents to come and experience the fun of planting, while the public was led to participate in creating and protecting the community environment.

Jumping from the bustle of city life to the rural practice and leisure activities in the bamboo grove, the project's main concept shows a desire of returning to the countryside and being away from the maddening crowd. This spiritual return creates a link between nature and the city. It is a modern Chinese garden that not only brings serenity, but also witnesses and commemorates the original natural environment of the city of Shenzhen.

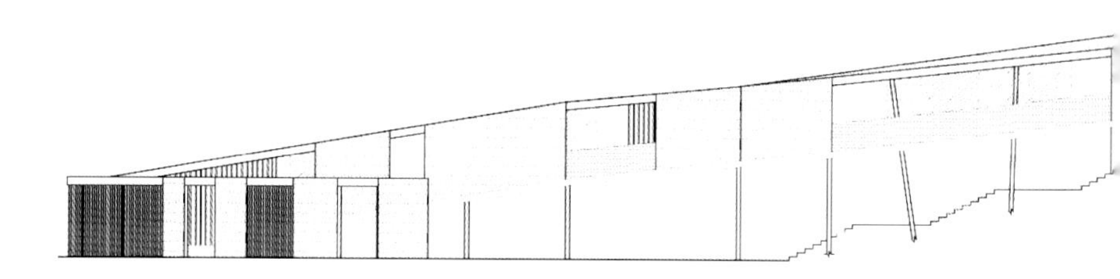

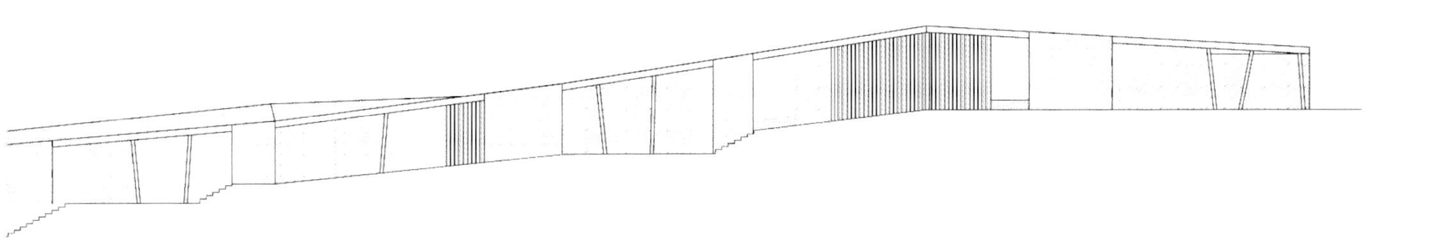

东门摄影广场 深圳

Dongmen Photography Plaza, Shenzhen

设计时间 / Design：2007 ~ 2008
建成时间 / Completion：2009
用地面积 / Site Area：4000m²
项目组 / Design Team：孟岩 | 朱加林 | 左雷 涂江 姚晓微 | 夏淼 张博 张震 罗琦
合作 / Collaborators：筑诚时代建筑设计（施工图） 黄扬设计（平面）
业主 / Client：深圳市罗湖区建筑工务局
摄影 / Photographer：孟岩 左雷 吴其伟

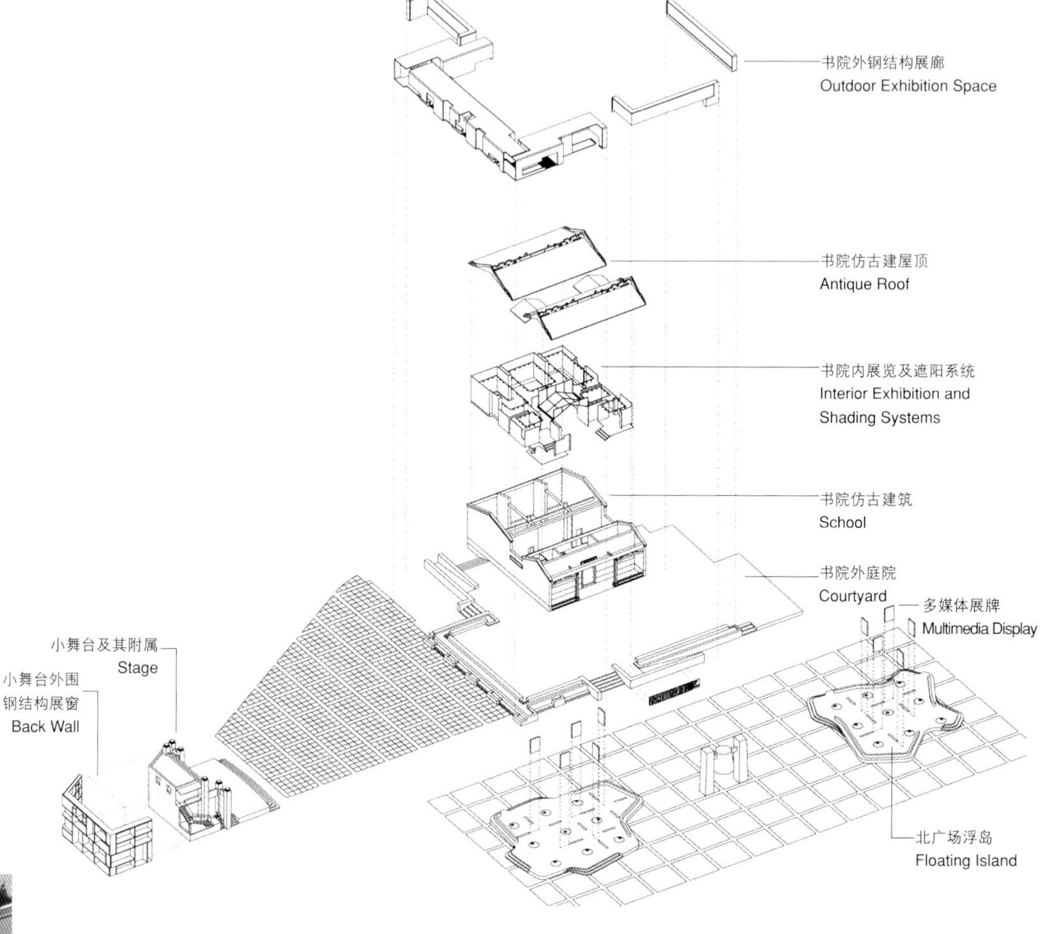

从一个小渔村的市集中心，到今天国际化大都市的两大中心商业圈之一，位于罗湖区的东门一直是深圳历史上最活跃的商业中心。设计任务是将位于其核心位置的原东门小广场改造为以摄影为主题的文化广场。这里有描绘古老市集的浮雕壁画、仿古大挂钟、空置的书院和以仿古照壁作背景的小舞台，还有些动物雕像和小天使，支离破碎地淹没在周围商城密布的嘈杂环境中。

设计的核心问题在于如何创造既能保留深圳仅存的历史片断、又能服务于当前人群休闲活动的空间。首先，上述记录历史的元素及现存树木花池被完好保留；书院加以翻新改造成为展示空间，加建一个“L”形室外展廊来展示影像，同时围合出一方形小院以衬托出书院宁静的气质。展廊与椰树丛中的两个不规则浮岛为人群提供遮荫、休息的空间，浮岛满布色彩浓烈的用瓷砖铺就的广东俚语图案。藏身展廊后、刻意保留下的长颈鹿、大白马也神气活现起来，令广场充满大众文化的气氛。高科技的多媒体的展示构思出现在建筑的外立面上，同时也是浮岛的构成元素之一，为这个叙述着“深圳故事”的摄影广场打上时代的印记。这个设计在传承着历史文化的城市空间中注入了充满活力的当代元素，也为新的“深圳故事”提供了最佳道具和舞台背景。

Dongmen in Luohu District has always been the most active business center in Shenzhen, from being a market center of a small fishing village in the 1980's to one of the two central commercial circles in today's international metropolis. This design task is to transform the former Dongmen square located in the core position of this area to a cultural square with photographic displays. On the site, there are scattered art installations and old building parts submerged obscurely within the chaotic environment such as a sculpted relief depicting the old bazaar, an antique large bell, a small old school, a stage with antique Zhaobi imitation as the backdrop, or even small animal statues and a small angel sculpture portrait.

The main concept is to create a space which not only retains the rare historical fragments of Shenzhen, but also serves the crowd for their leisure activities. All the elements above, as well as the existing trees and flower beds are well-retained. The school is renovated and transformed into a gallery. An outdoor exhibition space with image display is arranged in an L-shape to enclose a small courtyard while creating a quiet temperament for the school. The gallery and the two irregular islands that are placed underneath the coconut trees provide resting places with shade. The floating islands are paved with saturated colorful ceramic tiles with a carefully arranged pattern showing Cantonese slang. The existing animal statues are preserved behind the gallery, energized by the rejuvenated courtyard, filling the public square with a jubilant atmosphere. The concept of high-tech multi-media display is used in the wrapper of the existing façade of the building beside the stage, and also in the installation display on the floating islands that marks the square with the impression of a contemporary era. The design injects the vibrant elements mentioned above for a square that preserves the history and culture, and also offers the best props and stage for the new "Shenzhen story".

KFC
小 冷饮
市场天天新款折上折

文锦花园广场 深圳

Wenjin Garden Plaza, Shenzhen

设计时间 / Design：2011
建成时间 / Completion：2011
用地面积 / Site Area：4800m^2
项目组 / Design Team：孟岩｜林挺｜魏志姣 于晓兰 刘洁
业主 / Client：深圳市罗湖区城市管理局
摄影 / Photographer：吴其伟

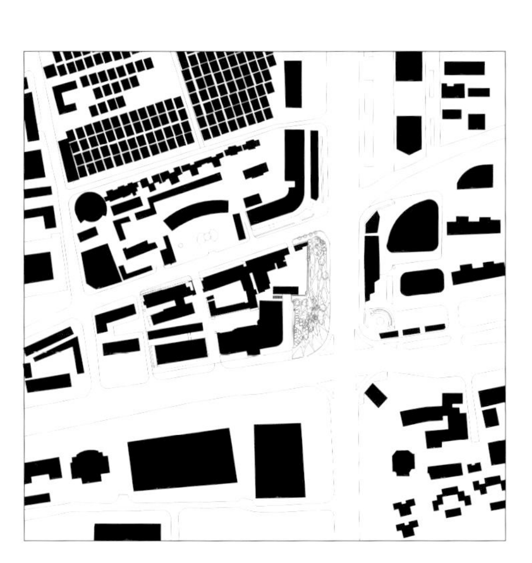

三十年来，从昔日小渔村脱胎换骨为今日的大都市，深圳的城市建设以平铺的方式快速向外拓展，而内部的城市公共空间并非都得到有效的利用和管理。

本广场位于深圳市罗湖区深南大道与文锦中路交会处，在过去的十年间，不断地有乔木被引入基地，但植物的种类与生长并没有得到良好的管理和维护。这个较大的场地在无公众监督或治安监督的情况下，成为消极的“避难所”。基地现状为我们提供了一个反思消极公共空间的机会：公共资源需要以一种平易近人的方式让公众享有平等的使用权利。为鼓励公众的参与，并建立场地平等使用权，同时创造一种介于绿地与公共空间之间、城市与自然之间模糊关系的介质，本设计策略制定为减少预先确定的刚性空间。在新的公共空间中，以不确定的方式让公众自主地形成使用习惯，同时也存在推翻已形成的习惯的可能性。因此，设计采用均质的、流动的形态与均等通行的道路系统。均质且具有方向感的形态基于场地植物位置的调整而产生，并依此形成一个通行系统。将南侧封闭的绿地空间转化为可通达的公共空间，激活该区域并吸引公众进入。不同尺度的休息平台与围合空间等手段的介入，提供了多样化的通行路径及丰富的空间体验。这个设计体现了都市实践一贯的策略：景观设计从一种作为背景的角色，变成触发城市故事的发生、激发城市活力的孵化器。

The past 30 years of economic development in Shenzhen has reformed what was a fishing village into the current international metropolis it is now; the urbanization progress has expanded with enormous speed and impact. However, during the process of such expansion, there are considerable amounts of public urban spaces being neglected, most of which lack of efficient management and proper inhabited planning.

This project is sited at the intersection of Shennan Road and Wenjinzhong Road in Luohu district, Shenzhen. The periphery of the site lacks of consistent maintenance, and is currently dominated by shrubs and bushes, therefore the enclosed site became problematic in terms of security management and public access. The condition of the site provided an opportunities to re-evaluate those left-behind public spaces. We intended to create an accessible public space. The priority of an urban environment is to enable the accessibility and sharing of land resources by different social groups.In order to introduce the equality of accessibility to the site, the project aims at blurring the boundary of urban green spaces and public activity spaces, further allowing the site to become a medium between natural and artificial parts of the city. The strategy is to diminish rigid urban spatial organization, and to redefine open spaces to involve potential more public activities of everyday life. To reflect the fluidity of the concept, the design seeks for a relatively even and fluent landscape form, constructing a convenient public network with clear directional routes and circulation system based on the existing vegetation pattern on site. The south section is activated by properly maintained green public spaces, attracting public interactions through interesting spatial experiences such as various scales of resting platforms and enclosed squares.

雅昌艺术馆　深圳

Artron Art Center, Shenzhen

设计时间 / Design：2009 ~ 2012

建成时间 / Completion：2010 至今

用地面积 / Site Area：12535m²

建筑面积 / Building Area：41504m²

项目组 / Design Team：孟岩 | 姚殿斌 | 周娅琳 饶恩辰 | 魏志姣 | 熊嘉伟

梁广发 吴春英 艾芸 孙艳花 黄志毅 王彦峰 | 林挺 陈丹平

合作 / Collaborators：Wendell Burnette Architects（室内）

极尚建筑装饰设计（室内）

广州容柏生建筑结构设计（结构）

深圳天宇机电工程设计（机电）

易科建筑幕墙

业主 / Client：深圳雅昌彩色印刷有限公司

摄影 / Photographer：吴其伟

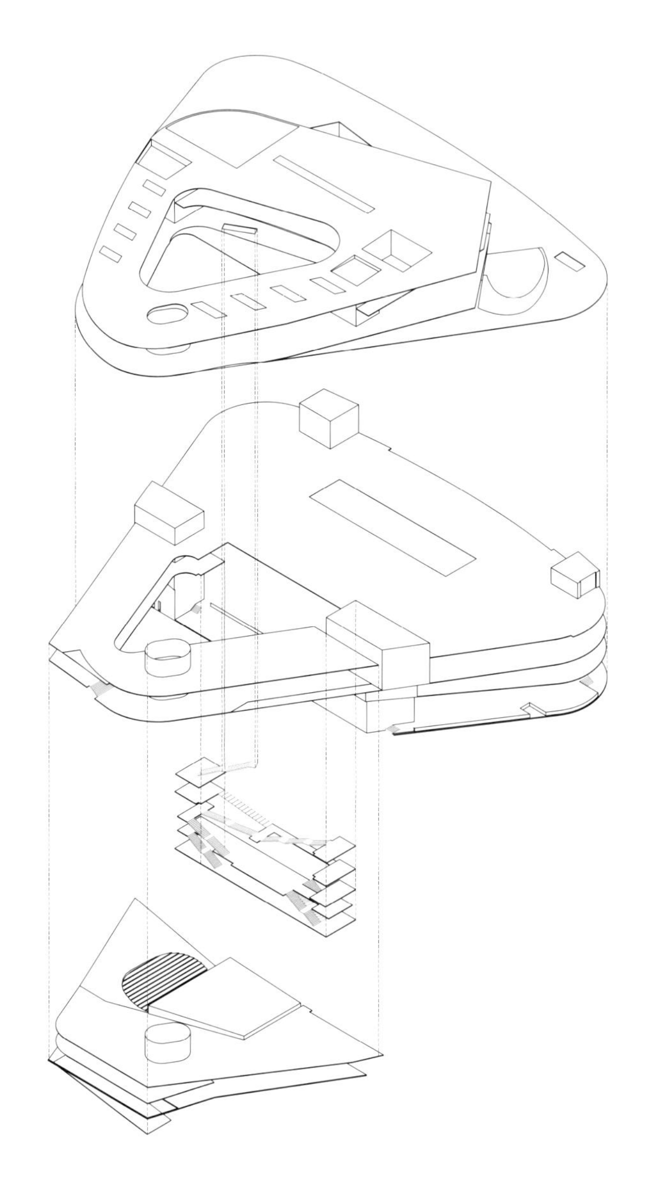

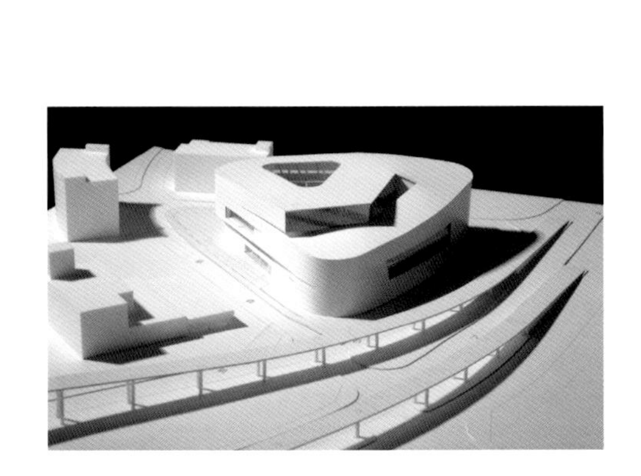

雅昌不仅拥有世界顶级的印刷技术，还拥有国内顶尖的雅昌艺术网——一个为艺术家建立资料档案的信息平台。基于这样的企业文化背景，加之远离城市中心的基地位置，拟建的深圳雅昌艺术馆被设想为一站式服务的印刷、文化综合服务中心。

基地位于被三条高速公路包围的城市边缘，周边的建设环境依然处于规划过程当中，具有相当的不定性。为了不被这样嘈杂、混乱的环境淹没，建筑注定需要率先成为定义这个区域的标志物。毗邻的三条高速公路，决定了建筑形体应以一种完整和连续的姿态，与大尺度的城市基础设施之间形成对话。将建筑体量做整的同时，又必须考虑如何消解巨大的体量。一方面，在保证完整和连续表面的前提下，裂解出实体面之间精致的开阖关系，为地面行人创造近人尺度的视觉感受；另一方面，在基地一端退让出一块三角形区域，改善周边的城市公共环境，从这个视角，建筑在每个侧面的表情各不相同。

在建筑内部，盘旋环绕的建筑形体围合出一个由美术馆、印刷工厂、办公空间、底层公共空间等多个部门共享的空中花园。作为独立于外部不利因素的企业内部环境，在空间组织方面，将艺术馆从公共属性的功能部分分离，与企业总部并列，独立地悬浮在景观资源最好的顶部，形成空间效果独特的独立艺术空间。这样的处理与流线的安排紧密相关，最终形成了能组织不同人群、不同参观方式的具有多选择可能性的内部丰富流线。

Artron is the foremost printing enterprise in China. As the core project of Artron's printing culture industry, the Artron art website is actively setting up to create archives for artists. Moreover, since the site is at the periphery of the city, the art center will need to be conceived with a one-stop multi-service printing facility and cultural center.

The site of Shenzhen's Artron Art Center is located on the city edge, surrounded by three highways. In order to keep away from the noisy and chaotic environment, the building is idealized to be a landmark to define this area. Since the main view of this building will come from the three adjacent highways, the building shape should be continuous and integrated with the large-scale urban infrastructure to form a dialogue between the two. Taking into account the volume as a whole, the architects must think about how to digest this huge volume. On the one hand, on the premise of integrity and continuity, the gaps between the different parts of the volumes were created to bring people the comfortable visual impression when walking in this territory. On the other hand, a triangular plot was reserved as a public park on the corner of the site to improve the urban environment. Observing from this small park, each side of this building is different. For the inner space of this building, the wreathed volume encloses an inner sky garden connecting art center, printing factory, office, the ground public space and is shared by the different departments. This inner sky garden blocks the noise from the outside.

For the internal organization of space, the art gallery was isolated from the public functions on the ground floor and placed in parallel with the corporate headquarters. Elevating those two parts to the top floor, the art museum has the chance to create the effect of an independent art space. The big move created opportunities to provide multiple choices for traveling in different routes to create a rich internal experience.

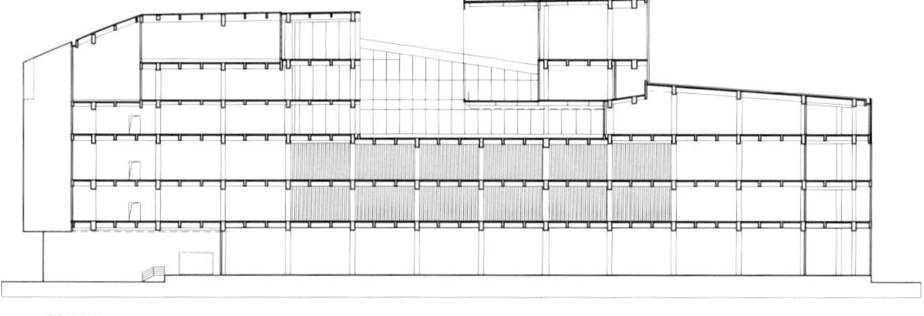

a-a 剖面图

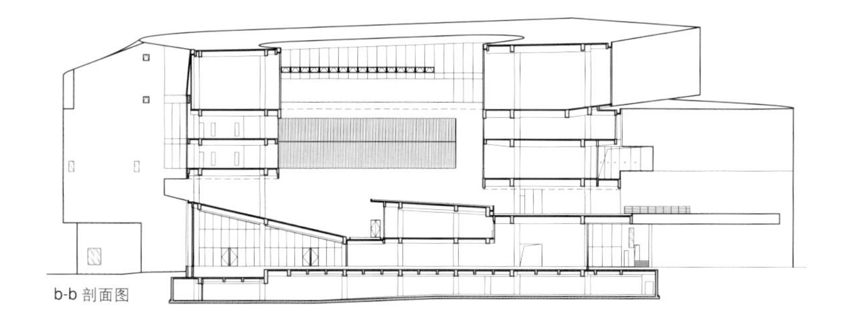

b-b 剖面图

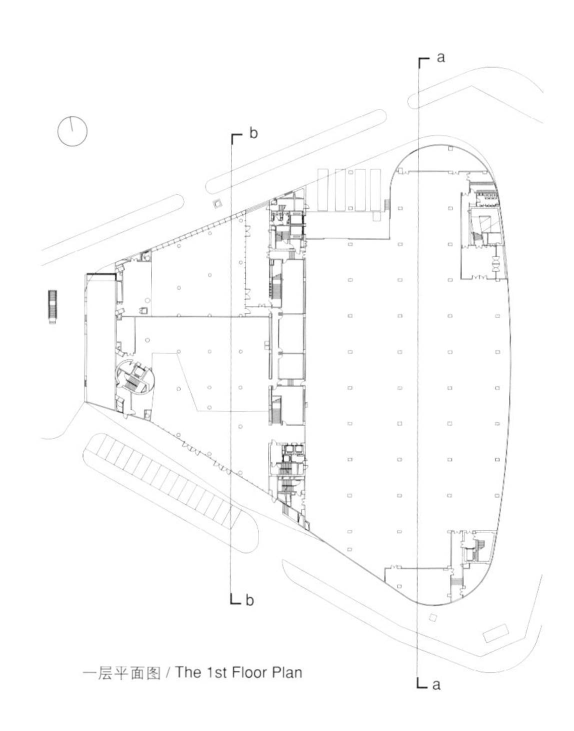

一层平面图 / The 1st Floor Plan

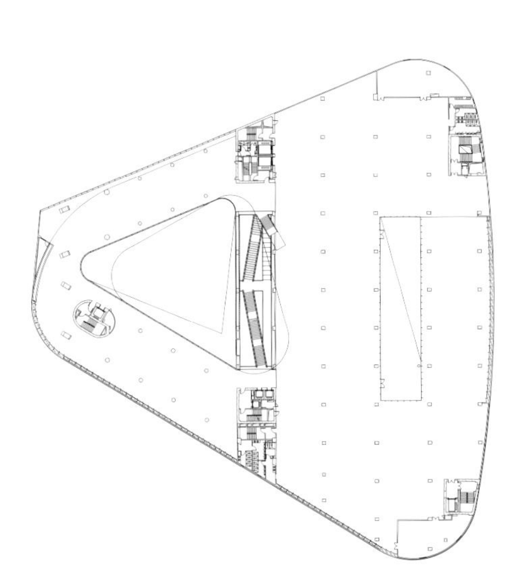

四层平面图 / The 4th Floor Plan

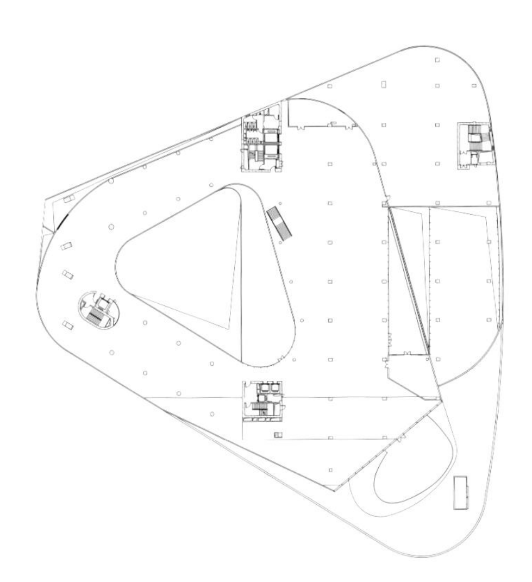

五层平面图 / The 5th Floor Plan

中电综合楼　深圳

Zhongdian Complex, Shenzhen

设计时间 / Design：2007 ~ 2012
建成时间 / Completion：2010 至今
用地面积 / Site Area：15292m²
建筑面积 / Building Area：41000m²
项目组 / Design Team：孟岩 刘晓都 | 姚殿斌 | 饶恩辰
陶剑坤 张震 涂江 | 姚晓微 李强 尹毓俊
合作 / Collaborators：中汇建筑设计（施工图）
广州容柏生建筑结构设计（结构）
瑞华建设（幕墙）
业主 / Client：深圳中电投资股份有限公司
摄影 / Photographer：孟岩 饶恩辰

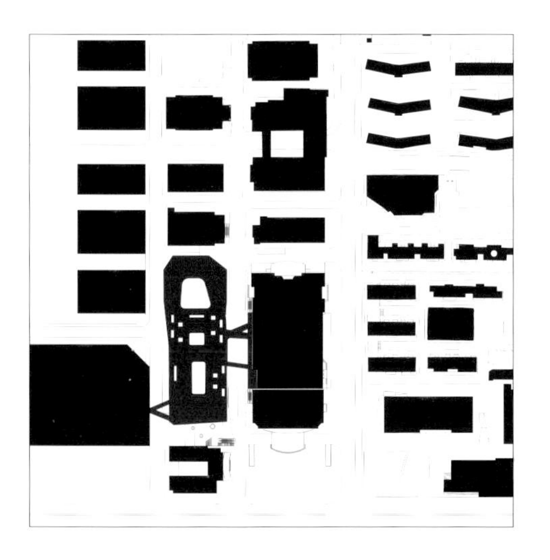

华强北是深圳最具活力的商业区，然而目前繁华的背后蕴藏着巨大的危机。人行、车行交通混乱，配套基础设施不足，使华强北由单一的电子配套市场向多功能综合商业街区转化过程中，遇到了巨大障碍。本项目用地位于华强北背街，是在多栋高层建筑的夹缝之中的一个每日约 20 万人穿越的基地。任务要求设计多层立体车库、商业、餐饮和旅馆等功能，更重要的是希望通过这座建筑重新梳理基地周边的人车交通系统，同时实现与周边的数栋商业建筑之间的立体相连。这样极高密度的城市环境蕴藏着巨大能量。

设计的挑战是如何在一栋同时容纳大量物流、车流和人流的巨大建筑中，在保证其各自系统高运转的同时，又制造新的商业机会和非同寻常的城市体验。于是一座"超级容器"应运而生，它包含了垂直叠加的底层物流集散场、沿街的商业及餐饮配套设施、中部的多层立体车库以及顶部的旅馆。场地内一座多层商业楼也被纳入到新的表皮下。建筑表面打孔金属板整体满足了多层车库的采光通风要求，建筑形体反映了在基地的空间限制下，试图挣脱这种限制所产生的扭曲，以及表皮形成的强烈的视觉效果。大楼中部被开凿出一个巨大的城市通廊，它既是过街装置，又通向商场的入口。此外原有街道穿越的方式也被保留了，并用多方向的坡道做了有次序地引导，使人流在穿越建筑物的同时能够方便地进入到各个商业楼层。巨大的采光井将自然光从高空引下，在通廊内弯卷的曲面上随时间而变化，每日几十万人流将从这里穿过，形成一种壮阔的城市图景。

Huaqiangbei, claimed to be the most vigorous commercial center of Shenzhen, is currently encountering an urban crisis: the chaos of pedestrian and vehicle circulation combined with the lack of urban infrastructure and facilities, which have become obstacles in its transformation from a simple electronic market to a mixed-use commercial district. The site is located in-between the main street of Huaqiangbei, in a void surrounded with overwhelming skyscrapers and criss-crossed by paths used by approximately 200,000 people per day. The client requested a multi-level parking structure with shops, restaurants and hotels. The more important issue, however, is to improve the circulation situation and to be more accessible to the nearby commercial buildings. Cities with such high density usually contain a great deal of energy. The challenge of the design is how to accommodate this amount of energy, such as movements of people, vehicles and other logistics into one large scale structure: ensuring the efficient operation of each system as well as creating new commercial opportunities and extraordinary urban experiences. Therefore, this "Super Capacitor" features as a hybrid building, consisting of a vertical distribution docking place at the lower level, commercial food services at the street level, a mechanical park system on the middle levels, and a hotel on the top floor. The existing five-story commercial building is preserved and wrapped by this new intervention. Entirely covered by perforated metal panels, the inner space of this "Super Capacitor" has adequate skylights and proper ventilation. The form of this hybridity, with this impressing façade, demonstrates spatial limitations from its surroundings and its struggle against them in the way of mediation. Horizontally, this building has been hollowed out and become a huge hallway, connecting the streets on both sides and the interior commercial entrance hall, maintaining the site-crossing which enriches back street life. Various connectors are offered for patrons to enter the commercial area in different levels. Vertically, two huge voids introduce skylights to this hallway achieving dramatic light effects on the façades.

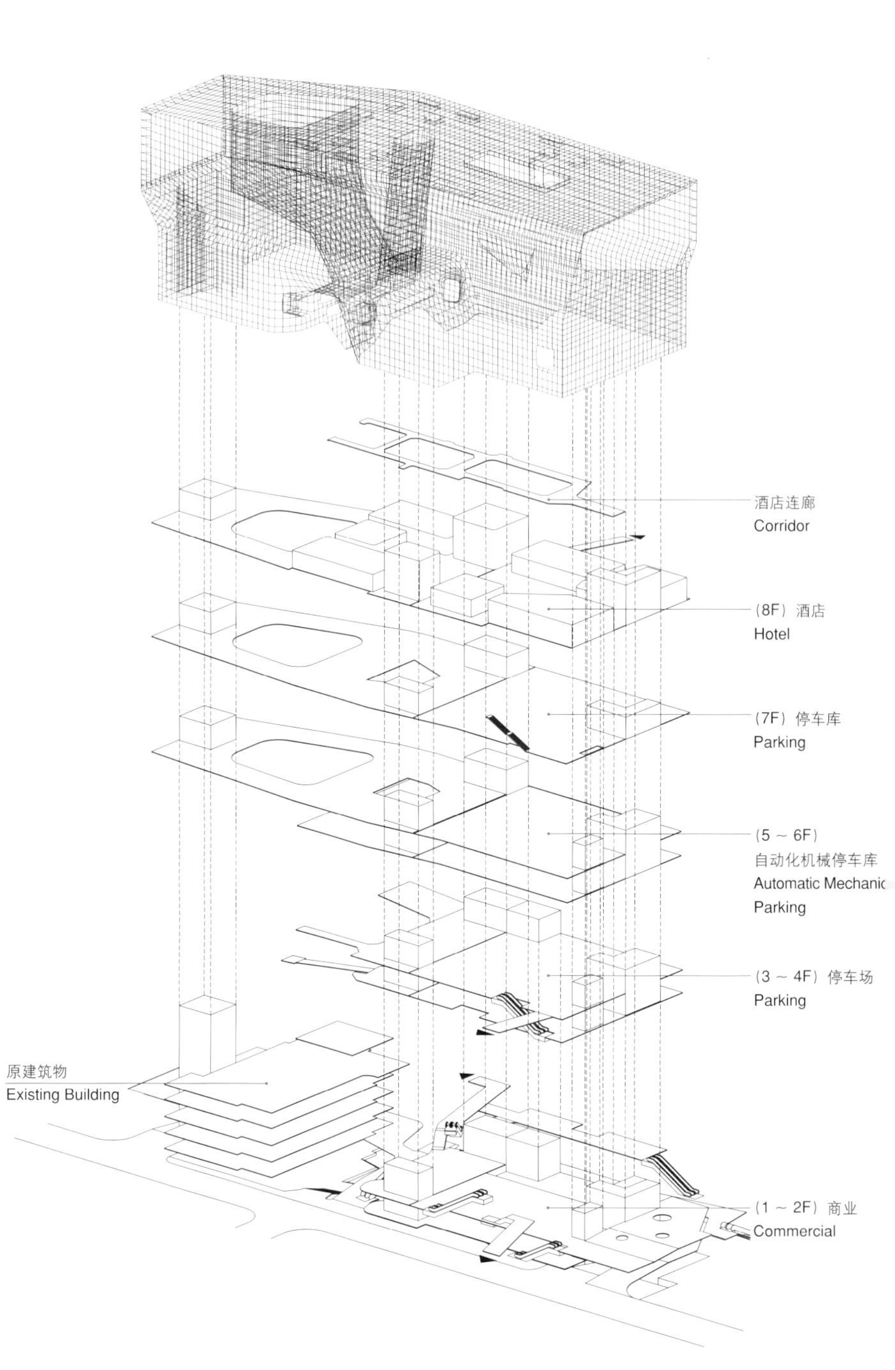

酒店连廊
Corridor
(8F) 酒店
Hotel
(7F) 停车库
Parking
(5 ~ 6F)
自动化机械停车库
Automatic Mechanic
Parking
(3 ~ 4F) 停车场
Parking
原建筑物
Existing Building
(1 ~ 2F) 商业
Commercial

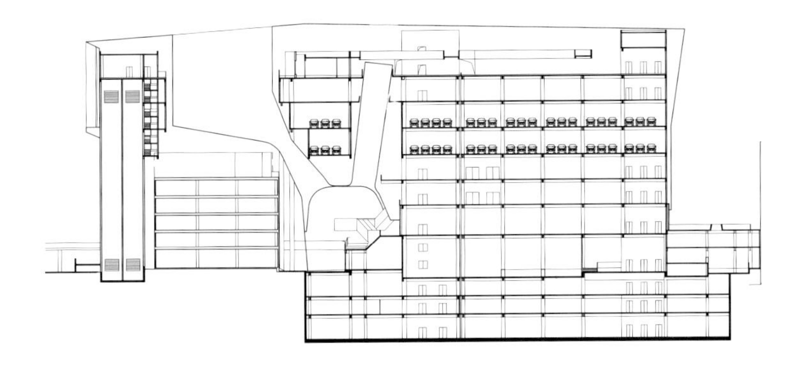

a-a 剖面图

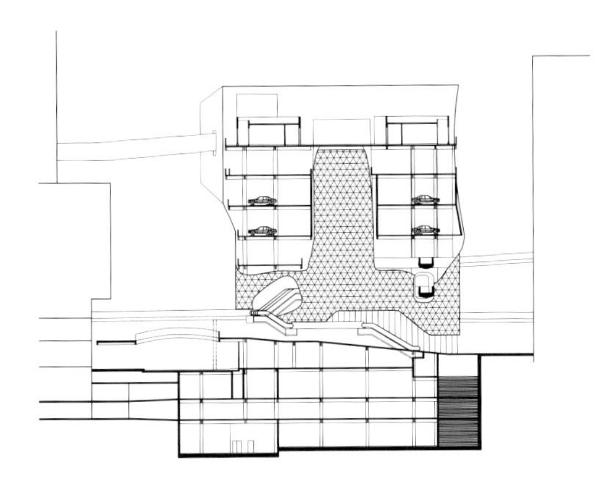

b-b 剖面图

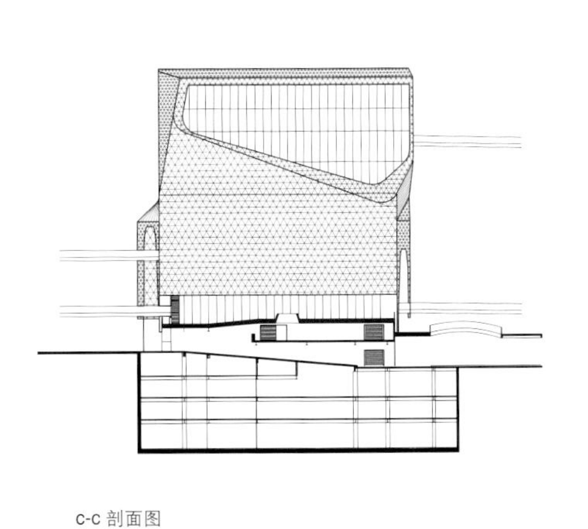

c-c 剖面图

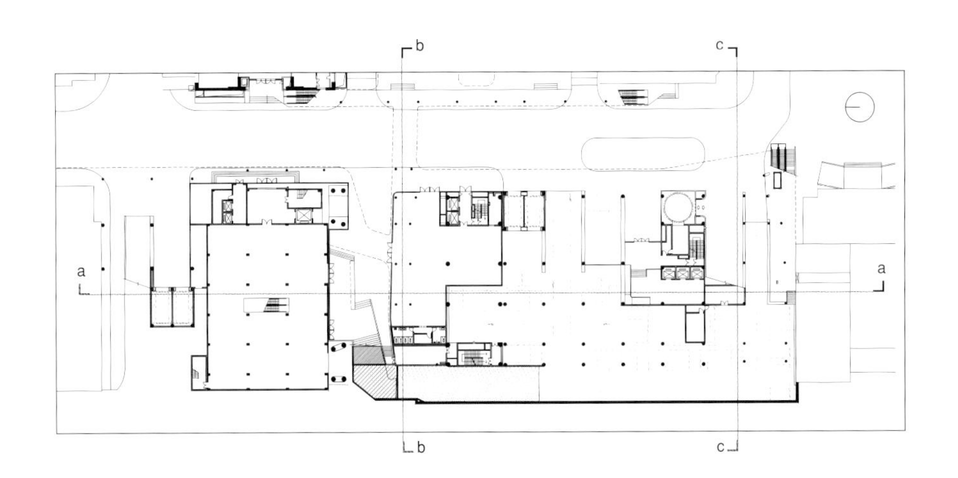

中广核大厦 深圳

CGN Headquarters Building, Shenzhen

设计时间 / Design：2008 至今
建成时间 / Completion：2010 至今
用地面积 / Site Area：10135m^2
建筑面积 / Building Area：158458m^2
项目组 / Design Team：孟岩 | 朱加林 徐罗以 吴文一 | 张震 罗仁钦
许小东 饶恩辰 | 李立德 吴春英 孙艳花 谢盛奋
合作 / Collaborators：广东省建筑设计研究院深圳分院（施工图）
广州容柏生建筑结构设计（钢结构）
柏诚工程技术（机电）
业主 / Client：中国广东核电集团有限公司
摄影 / Photographer：吴其伟

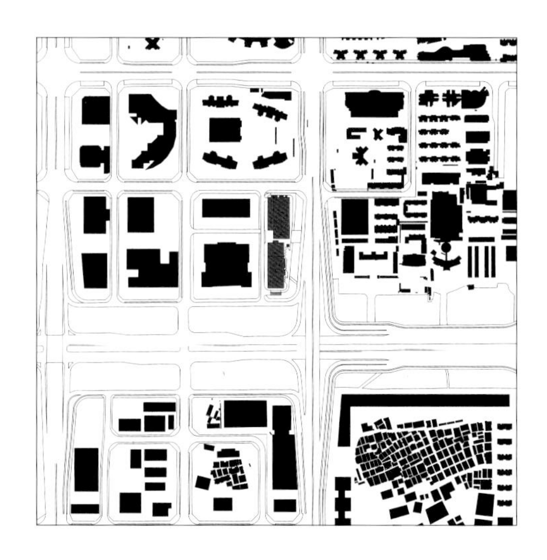

经过十余年的超速发展，深圳中心区已经初具规模。然而伴随着每座单体建筑对自身独特性的极力追求，建筑物之间缺乏敏感的呼应，城市中心建筑组群的完整性正在丧失。重归高层建筑的基本特点：高效舒适、节能环保、造型明晰，回归建筑经典，有时反能使建筑物在当今纷繁混杂的城市背景中脱颖而出。

用形体和空间组织来表达建筑在节制和内敛之中蕴藏的能量，使总部大厦远观有清晰简明的形体轮廓，近观又有丰富多变的肌理层次。两座大楼形体尽量向东侧基地充满，并且在平面和空间上相互交错，最大限度地利用东西两侧景观资源的同时，形成两楼咬合、互动的整体形象。

建筑立面肌理在简洁的体型之中传达数字化的美学特征，以单元化、标准化的重复以及变异，组合发展出一整套建筑立面语言体系。建筑表面弃用了大面积的玻璃幕墙的作法，回归到窗墙体系，塔楼窗洞在大小、方向、凹凸上的变化逐渐过渡到裙楼部分的网格裂变，并以此支撑起悬浮在空中的几组公共空间。匀质模块基础上的单元作渐变以及局部剧变这种空间构成体系暗示了核电作为未来能源支柱产业的形象特征。夜晚，灯光透过变化的格网呈现出晶体般的表面，整个建筑外部又似乎成了一个可以容纳无限变化的显示屏。外墙采用金属质感的较深沉的色彩，进一步强化了大型企业总部的气度。整个设计展现出一个国际化企业所具有的严谨、稳健、前瞻的形象特征。

After a decade of speedy developments, the structure of Shenzhen's Central Business District (CBD) has begun to stabilize. However, every individual building is pursuing its own uniqueness and lacks sensitivity between urban dialogues. Therefore the urban center is losing its totality. Returning to the basic problem of high-rise buildings—such as the use of energy saving technology, long-term sustainability, and clear and concise design approaches—how to make the design distinctive from the chaotic city centre and return to classic forms.

The architectural form and organization of space expresses the abstention from power and gives the CGN Headquarters a simple and concise figure from afar. The two tower blocks occupy the site eastwards, and interlock in plan and space. An image of two linked and interactive buildings is formed by fully utilizing the landscape at the East and West side. In this area, two blocks respond to each other and form a bracket in the air.

The simplicity of the building facade texture conveys digital aesthetics. Modular windows repeat and vary throughout the skin, and become the main architectural vocabulary system. These windows vary in size, direction and depth, which gradually transitions towards grid fissions and extensions, supporting the floating public spaces. Based on these modular units, the spatial system of this gradient changes and partial upheaval acts as a metaphor for the fact that nuclear power is becoming a major energy industry. At night, light travels through the grid, transforming the façade into a crystalline skin, and transferring the whole building into a screen that contains infinitive change. The dark metal façade emphasizes the corporal expression of CGN's headquarters. The design exhibits the preciseness and solidity of a well-known technological enterprise, echoing its ambition of becoming a more international and future driven industry leader.

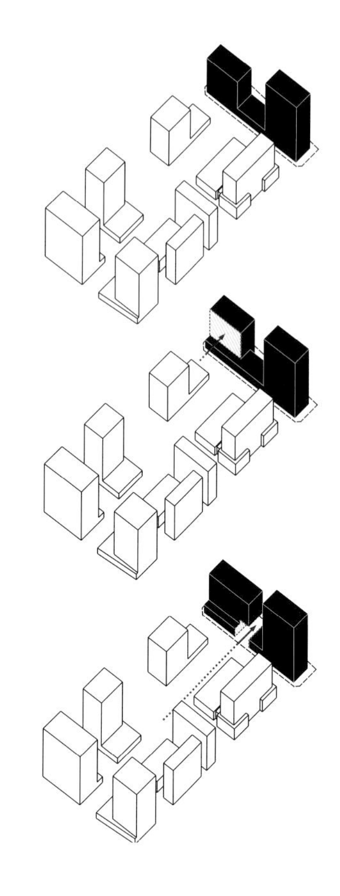

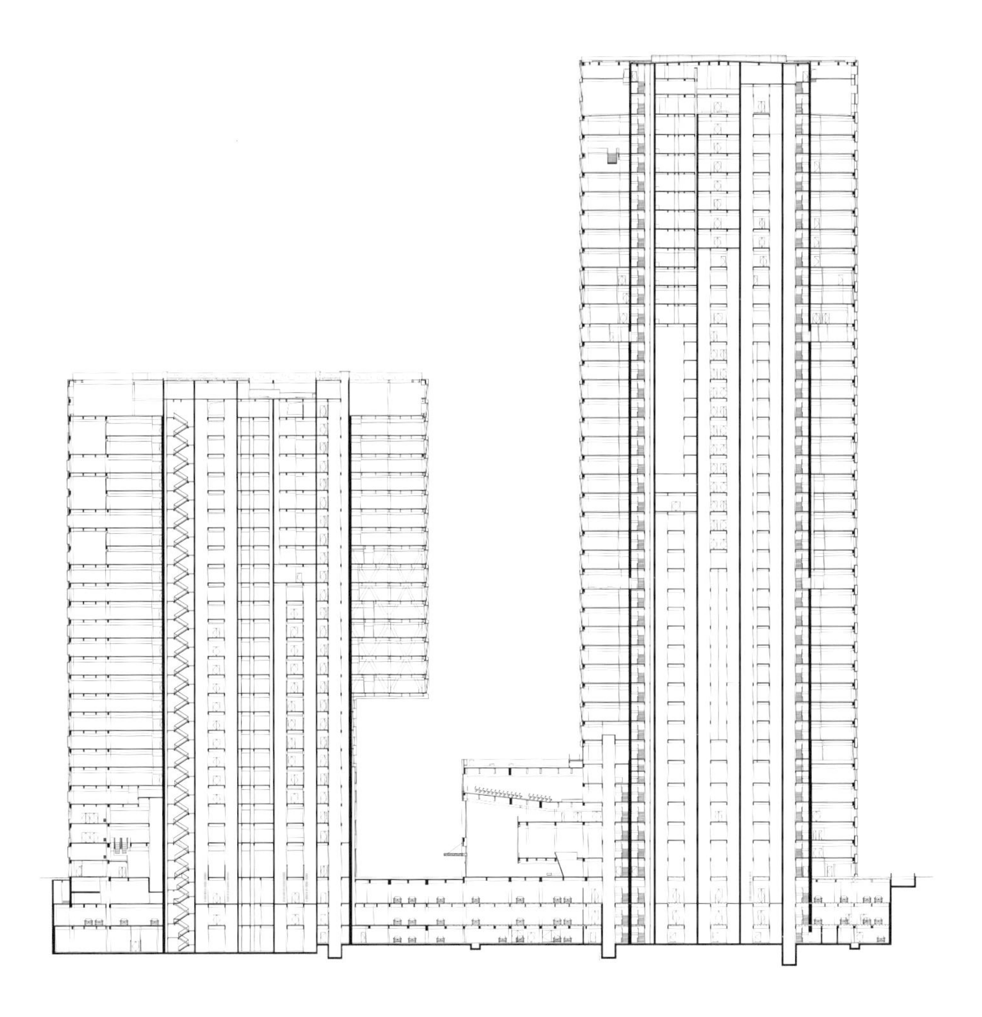

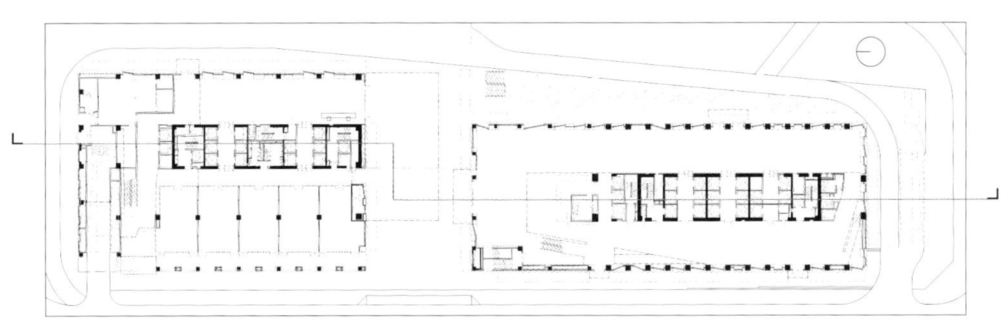

一层平面图 / The 1st Floor Plan

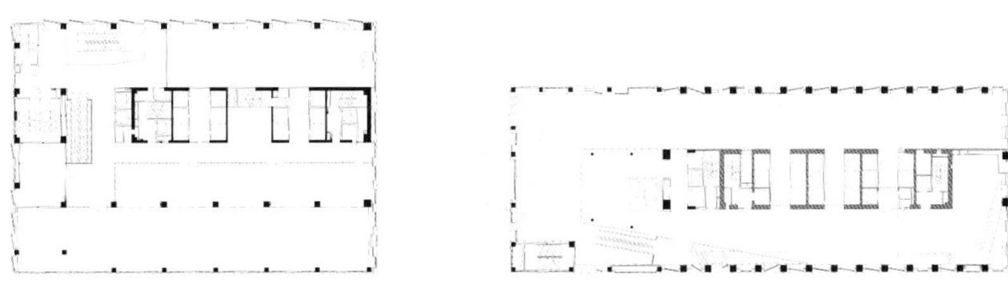

五层平面图 / The 5th Floor Plan

卓越时代广场二期　深圳

Excellence Group Times Plaza Phase II, Shenzhen

设计时间 / Design：2006 ~ 2010
建成时间 / Completion：2010
用地面积 / Site Area：9500m²
建筑面积 / Building Area：110000m²
项目组 / Design Team：孟岩 刘晓都 | 罗仁钦 张震 陶剑坤 郑颖 | 乔锴 陈文娜 李晖 黄煦 丁钰 张星 唐圆圆 | 邢果 黎靖 魏志姣 胡伊硕
合作 / Collaborator：深圳机械院建筑设计有限公司（施工图）
业主 / Client：卓越置业集团有限公司
深圳市祈年建业投资有限公司
摄影 / Photographer：吴其伟

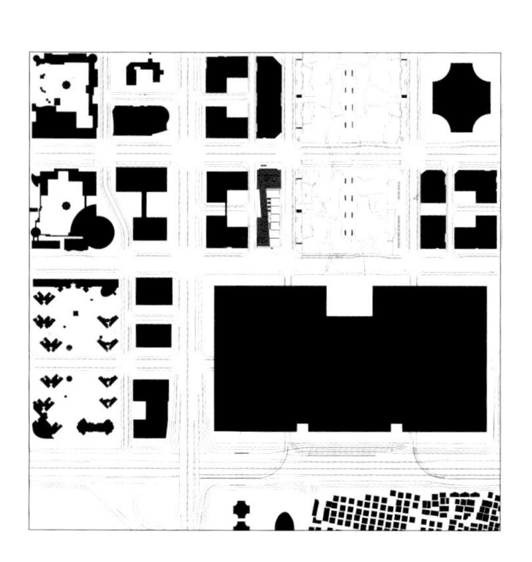

本项目是集高档商业、酒店和办公为一体的城市综合体，基地位于深圳繁华的福田中心区。建筑西侧背靠两座 200 米高的塔楼，再往西是规划中 350 米高的超高层塔楼；建筑东侧面向半下沉的中心商城及其上方宽阔但人迹罕至的屋顶花园。因此该项目是中心区西侧超高层塔楼群与中部大尺度水平蔓延的建筑体量之间的过渡。

建筑师采取一种积极回应场地特点的策略，说服规划部门将原本规划中两栋并列的方形塔楼从形态和空间上加以整合，裙楼与北侧的写字楼等宽形成垂直的"L"形体量，酒店部分的高层板楼与多层连接体形成另一个"L"形体量，让两个"L"形体量交错拉接形成一气呵成的整体。立面上，酒店倾斜的外表由错落通透的单元体幕墙叠合而成，如一片片鳞甲反射天光，并使纯粹的几何形体更加有力和具有很强的识别性；而办公部分采用竖向线条，以突出高耸挺拔的效果，并与酒店的体量呼应。充满基地的裙房中部两层贯通，连接东侧中心商城与西侧城市广场，从这里也有连接两侧建筑的二层连桥穿过。裙楼部分为综合性商业和餐饮设施，公众可以穿过裙楼的公共部分一直到达位于四层屋顶的坡地花园。裙楼屋顶南低北高，南端连接酒店的空中大堂，向北逐渐拾级而上与室内泳池等休闲设施相通。两主楼间的连接体具有独特的景观资源，且底层架空以将屋顶花园整体贯通。

城市综合体在新兴城市的发展中有更大的责任，它需要敏锐地把握和整合周边各种空间资源的潜力，为新的城市生活提供更加丰富多样的舞台。

The project is an urban complex containing a hotel, office, public spaces and high-end shopping malls. The site is located in the center of Futian CBD in Shenzhen, with two existing 200-meter towers to the west and another pair of 350-meter-high skyscrapers in the future development plan. The shopping plaza with its empty roof gardens is located to the east. Therefore the main concept for the project is inter-connecting media providing transition between the high-rise towers on the west and the horizontal large scale shopping center. The planning bureau was convinced to alternate the mass of the two planned cubical towers in the urban design guidelines, resulting an interlocked unity based on two L-shaped massings; one being the podium with the same width of the office tower on the north addressing the verticality and the other one being the podium of hotel constructed with multiple level connectors extruded from the hotel slabs.

The façade of the hotel, layered with transparent modules with various depth and strong geometric form, is slightly inclined to reflect natural lights from multiple angles. The office tower façade uses straight lines to amplify the verticality effect, while echoing the hotel's tower form. The podiums are divided into mixed-commercial zones and a food and beverage area, which are inter-connected with various sky-bridges at different levels, merging the pathways between the eastern shopping mall and the western city plaza. Public circulation can penetrate right through the four levels of the podium and reach the sloped roof garden, which has a natural inclination from north to south. The stepped zones are arrayed from the south gradually towards the north. The sky bridge sandwiched between the two towers has an excellent view with column-free overhead structures, linking the rooftops as an entity to enhance public activities. The project acts as an Urban Complex in an environment where fast modernization is occurring, which strongly emphasizes the efficient usage of sacred land resources, providing a rich and diverse stage for a new type of urban living style.

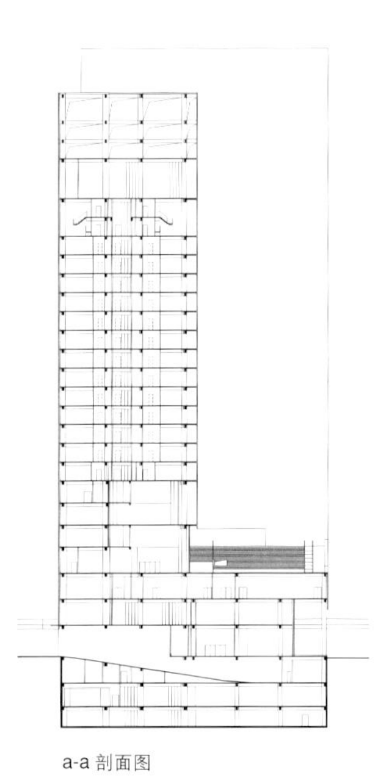

a-a 剖面图

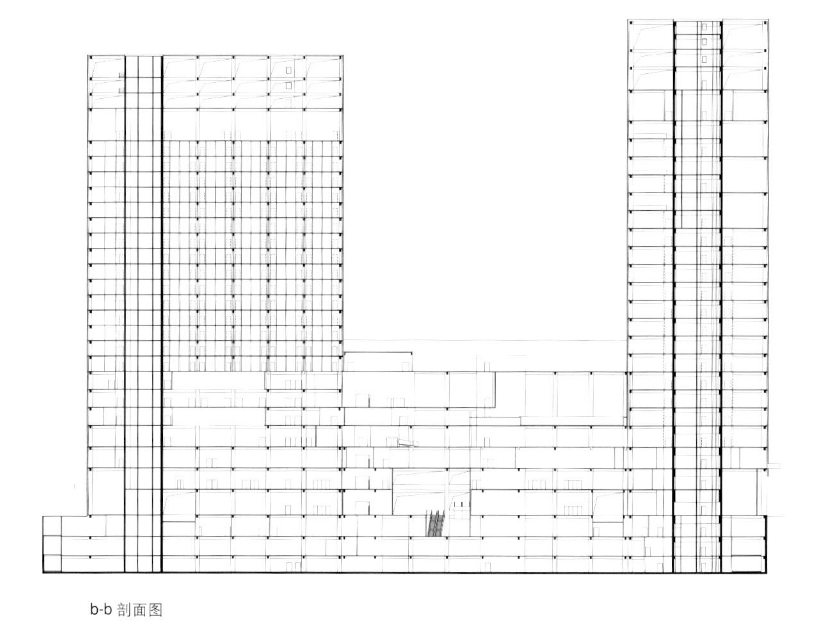

b-b 剖面图

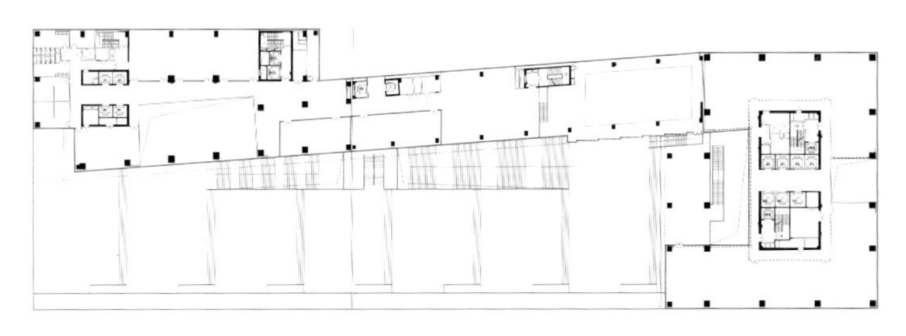

五层平面图 / The 5th Floor Plan

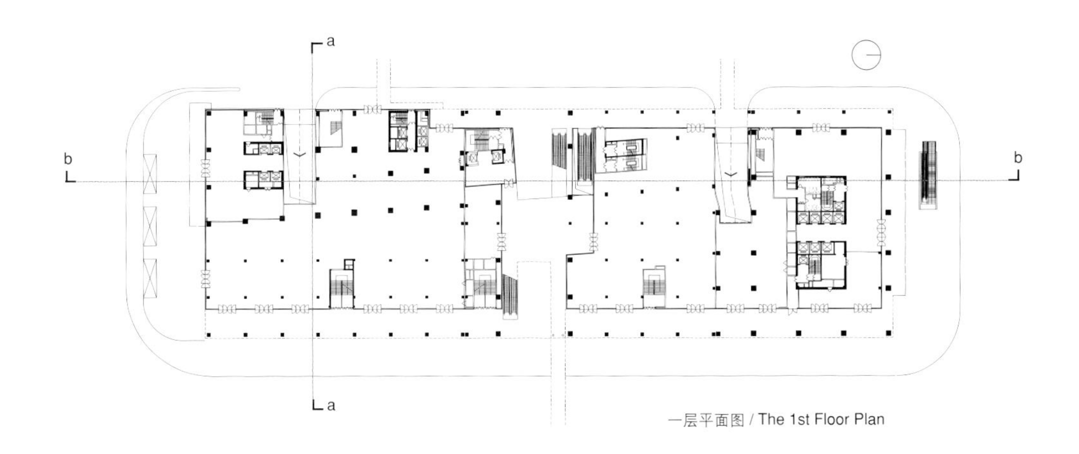

一层平面图 / The 1st Floor Plan

大中华 IFC 深圳

Greater China IFC, Shenzhen

设计时间 / Design：2006 ~ 2010
建成时间 / Completion：2011
用地面积 / Site Area：18600m²
建筑面积 / Building Area：197801m²
项目组 / Design Team：刘晓都 孟岩 | 罗仁钦 傅卓恒 林海滨 | 李达
陈耀光 乔锴 黄煦 李晖 沈艳丹 尹毓俊 陈文娜 张星 王雅娟
梁广发 张新峰 熊嘉伟 熊慧晶 Mathias Wolff 刘浏
合作 / Collaborators：泛华工程有限公司（施工图）
业主 / Client：大中华国际实业（深圳）有限公司
摄影 / Photographer：吴其伟

一直处于持续高速发展的深圳，是一个带形城市。若要迅速地阅读这个城市，沿某一主干道驱车行驶上一段时间，整个城市的剖面便可尽收。作为深圳城市主动脉的深南大道，便是这样一条可以快速阅读深圳的剖切线。深南大道南侧的大中华 IFC 项目，位于深圳中心区东侧，被城市快速干道所环绕，属于"快速通过区"。人们对这一区域的印象都是在"快速通过"之中获得的。在这样的"快速通过区"，建筑需要一个富有感召力和公众识别性的形象。

这一项目用地的最大特点是狭长，它整整占据了300余米长的一整个街区。在深南大道，尤其是位于最有视觉吸引力的福田中心区，这种地段气势是绝无仅有的。因此，设计着力于利用这一沿街面塑造连续的形体，将四个单体建筑转化为一个即使在高速行进中也可以体验到的连续的建筑体量。酒店、办公楼、商务公寓和大型商业的造型都融合在一条连续的折叠面上。这一长而完整的带形体量，又很好地隐藏了南侧杂乱的城中村。只有 24 米进深的基地，使其可以轻易地将活跃的城中村商业活动和新大楼的底层商业融合起来，使城中村的活力能够渗透到深南路上。这样的一体化的建筑形式，也是探讨复合型城市综合体的一个新的表达方式。

Shenzhen, a linear-shaped city, has been seen rapid growth in the past few decades. Simply by driving along one of its main arteries, such as Shennan Avenue, one can get a clear and sectional impression of the city's urbanization achievements.

The Greater China IFC is located along the south side of Shennan Avenue, on the east side of the city's central district. It is surrounded by high-speed city avenues, and as such requires a recognizable landmark. The most important characteristic of this site is that it is long and narrow, occupying a whole block stretching some 300 meters. Such a scale for a site along Shennan Avenue and within the city center is quite unique.

Based on these site characteristics, the architects aimed to design a series of continuous buildings along the street, and to transform four individual buildings into a unified series that can be identified even in a fast-moving vehicle. Hotels, offices, business-apartments, and shopping areas are all integrated within the undulating and folding façade. In addition, this belt helps conceal the chaotic urbanism of a village in the city to the south. Only 24 meters deep, the base elegantly combines the dynamic businesses of the villages with retail space in the new buildings, thus injecting the vitality of the village to Shennan Avenue.

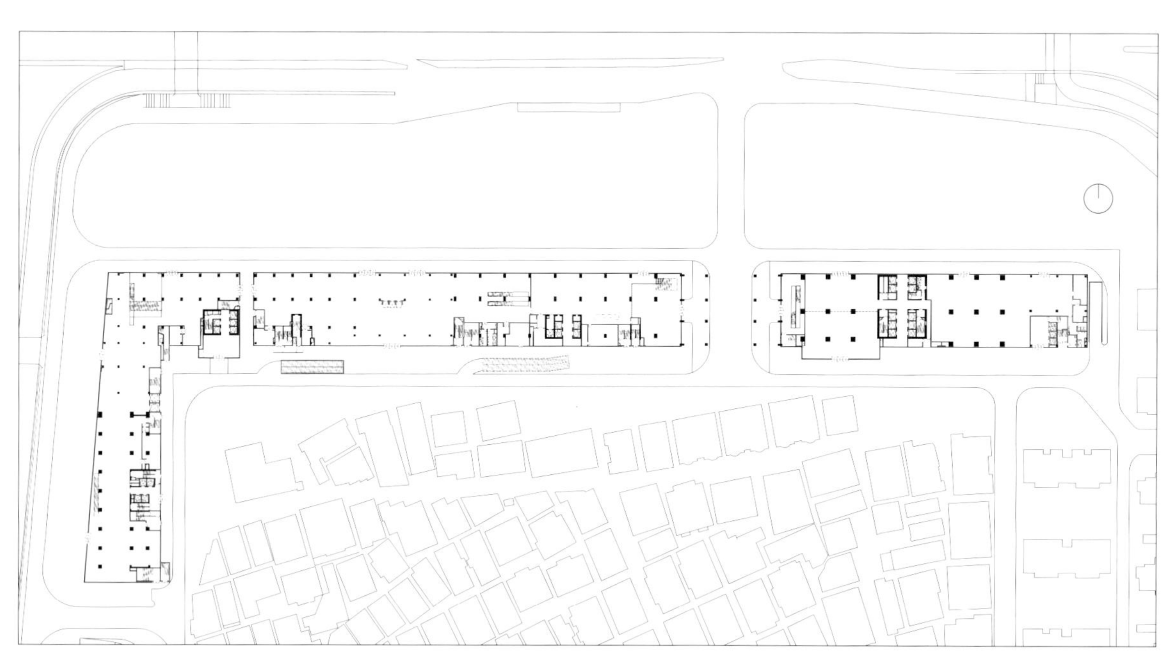

一层平面图 / The 1st Floor Plan

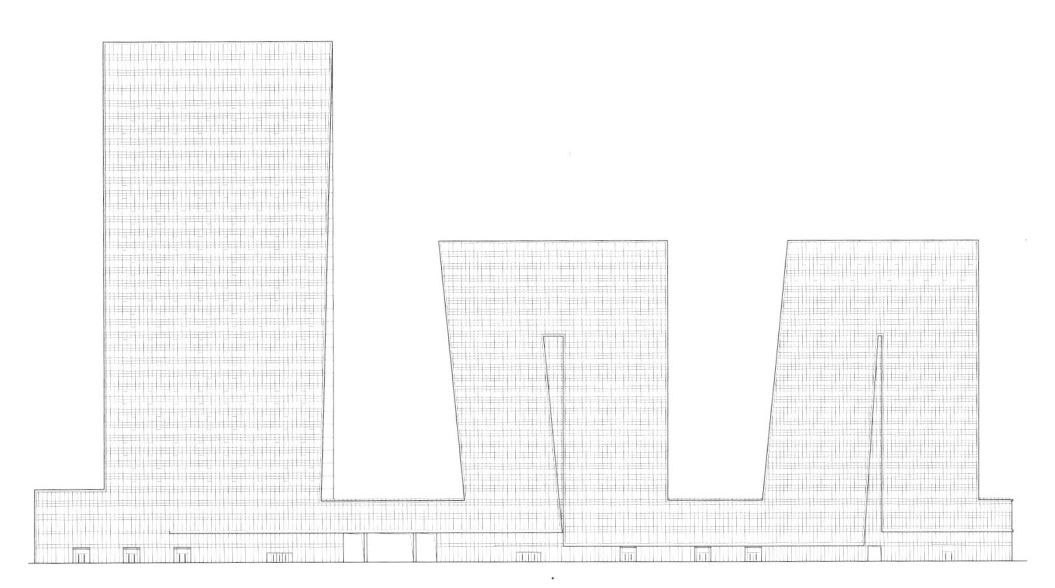

北立面图 / North Elevation

北京 CBD 核心区概念性规划及单体设计

Beijing CBD New Core Urban and Building Design

设计时间 / Design：2010

用地面积 / Site Area：159600m²

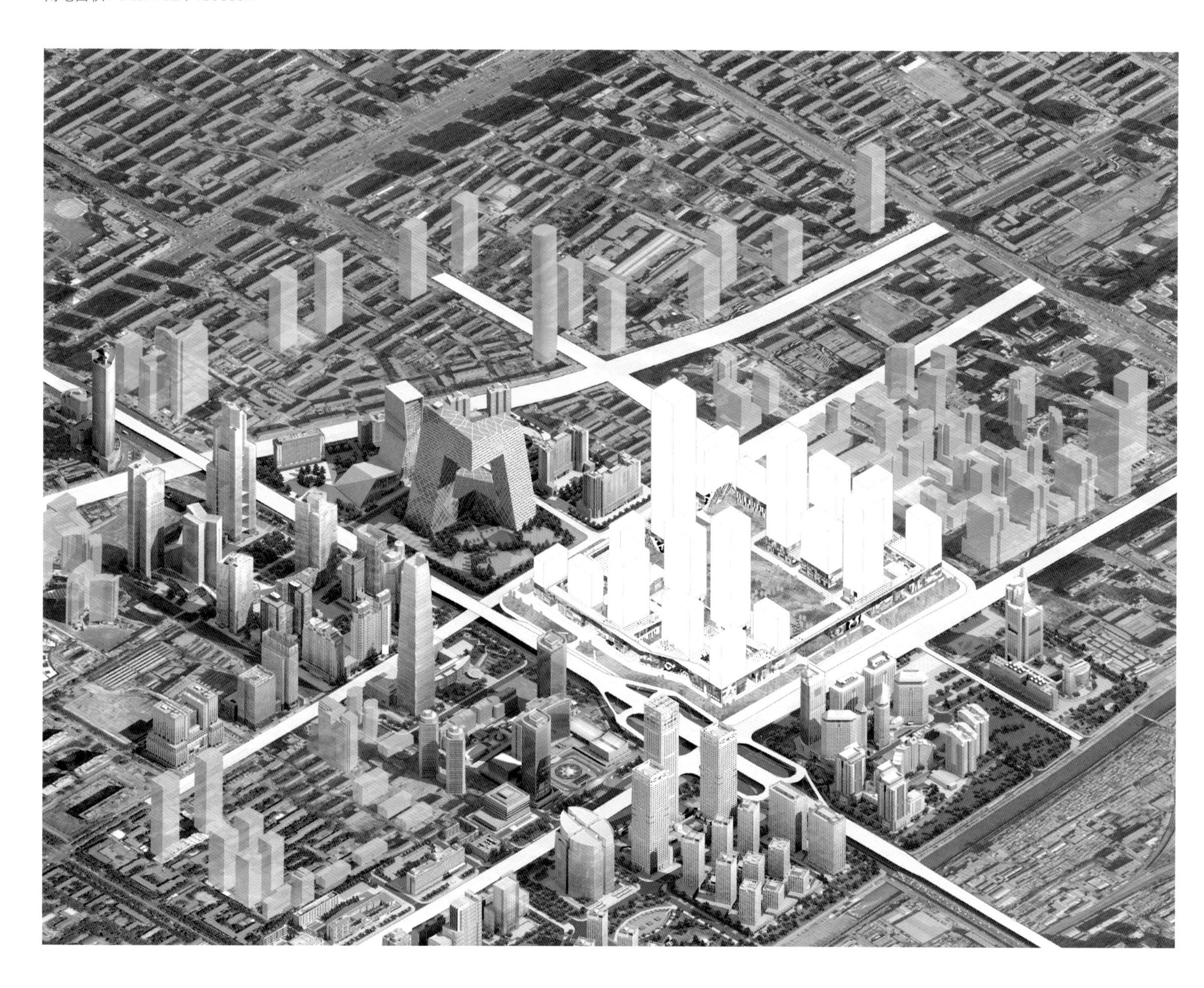

作为首都门户的北京 CBD 核心区的规划，是在全新的社会技术、经济、文化条件下探索全新的摩天楼街区模式的一个机会。有别于千篇一律的棋盘式的城市高层核心区，这个复合的城市中心将汇集当代在城市基础设施、交通、能源、功能、景观、管理等诸多领域的认知与成就，设计低碳时代的未来城市。我们的提案没有侧重于塑造高层部分乌托邦式的单体造型，而是利用 CBD 建筑的集群优势，整合路面空间、地下空间和裙房空间的资源，形成一个网络化的城市公共空间体系，并赋予这个体系以鲜明的视觉文化特征。

方案一是基于北京城的历史文脉以及 CBD 核心区的相对独立性，将外围各个单体的建筑裙房沿 CBD 的外轮廓连接，形成围合式的城中城意象；而城内各建筑的公共空间相互开放连结成网，资源高度整合，设施可共建、共享和互换，并与城外连通。北京的历史印记是紫禁城（FORBIDDEN CITY），而 CBD 核心区的这一“开放之城”（UNFORBIDDEN CITY）则将北京推向未来。

方案二是把城市从僵硬的几何体系中解放出来，用更自然的山体形式和地貌形态，形成一个“幻境之城”（DREAMLAND CITY）：高层建筑被山峦般的裙房联结成一个体系，地下空间和中央公园形成一个个异质功能组成的浮岛，因而构成一个趣味多元而形象统一的立体城市。这两个方案在提高复合型城市中心的效率同时，都致力于在公共空间中植入主题性的文化设施，并以此推动新的文化产业的发展，从而使城市的建设成为推动当代文明发展的主动力。

Scheme 1

Scheme 2

Beijing, the capital of China, is proposing a future CBD core urban plan, with a foundation of new social technology, economic status and cultural reform, and exploring opportunities of re-defined street-blocks with skyscrapers. Different from the generic grid layout of urban high-rise CBDs, this particular mixed-use urban complex would aim at seeking a future city model, with collaboration of experts of infrastructure services, traffic interchanges, energy saving, landscape design, collective urban management and sustainable development. Our proposal does not emphasize the normally utopia-like singular entities of skyscraper design. It has efficiently incorporated the collective advantages, with integration of ground level and underground spaces together with the podiums to form an invisible network of urban public infrastructure system. The system then acts as a backdrop of further articulation of visual cultural characteristics.

Scheme One is based on the historical background of Beijing city and reflects the CBD area's relative independency, creating an enclosed impression of "city within city" by connecting the fragmented building podiums around the outer periphery of the CBD. Within this periphery the public spaces are opened up to form an inter-connected information and resource network with excellent sharing efficiency. Beijing's historical imprint is the iconic Forbidden City, while the future development of CBD implies an "Unforbidden City". Scheme Two is to liberate the city from the existing grid fabric, the podium would flow in a form of natural topography like a dreamland built upon mountains. Towers are connected with floating islands, resulting in a unified image of a three dimensional urban life concept.

The two concepts both amplify the theme of public services with cultural engagement, are guided by the highly efficient CBD infrastructure, and act as a major driving force of contemporary urbanization.

Scheme 1

北京互动新媒体文化中心

Beijing Interactive New Media Building Center

建筑面积 / Building Area：79045m²

项目组 / Design Team：孟岩 刘晓都 | 林怡琳

饶恩辰 张震 | 曾冠生 陈兰生 林挺

Manuel Sanchez-Vera Michael H. Rogers

Mares E. C. Marc Deidda Alessandra

李耀宗 黄艺宏 汪源 马晓瑛 朱伶俐 陈斯

业主 / Client：盛大 / 阳光传媒

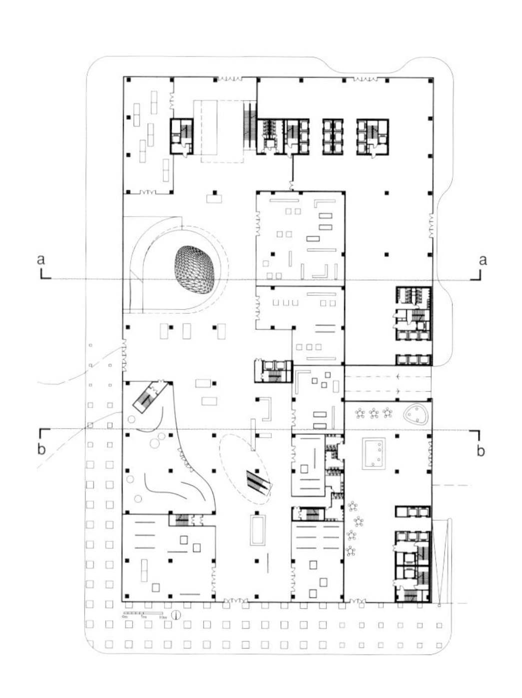

一层平面图 / The 1st Floor Plan

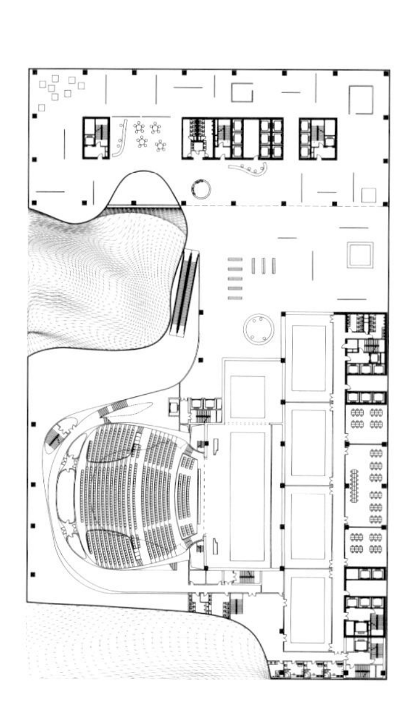

数字媒体博物馆 / Digital Media Museum（The 7th Floor Plan）

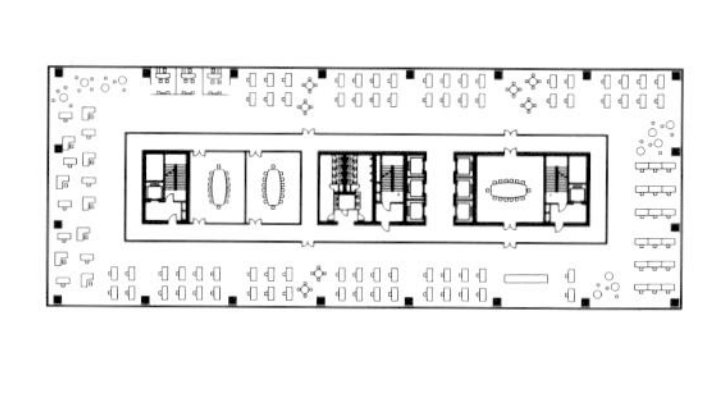

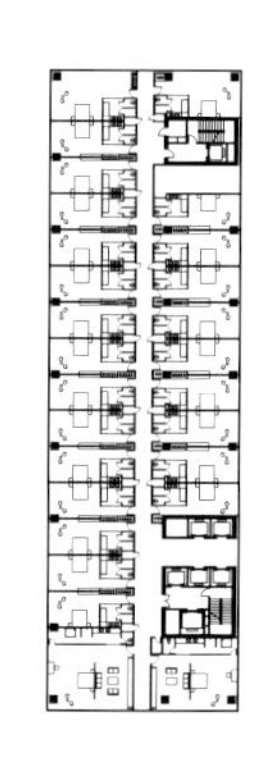

高区标准平面图 / Upper Typical Floor Plan

a-a 剖面图

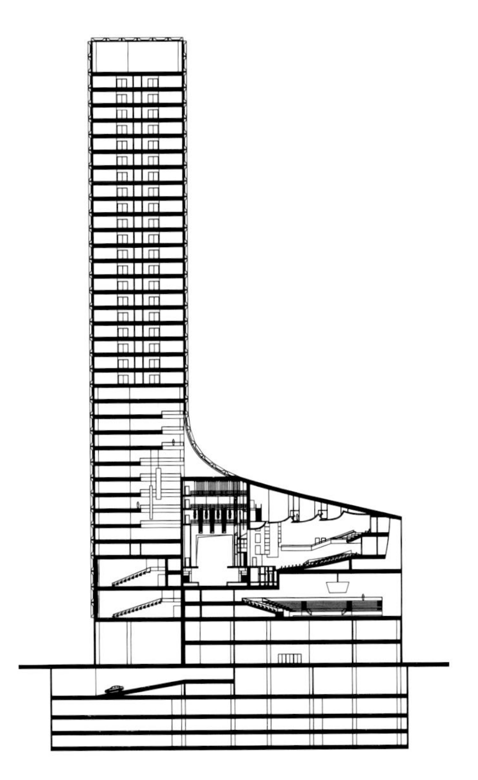

b-b 剖面图

Scheme 2

中国宝马总部

China BMW Headquarters (CBD)

建筑面积 / Building Area：132000m²

项目组 / Design Team：王辉 吴文一 | 郝钢 | 杨楠

林秀清 Jennife H. Ha Deborah A. Richards

合作 / Collaborators：北京嘉工国际建筑（结构）

何新城事务所（城市设计）

业主 / Client：阳光新业地产股份有限公司

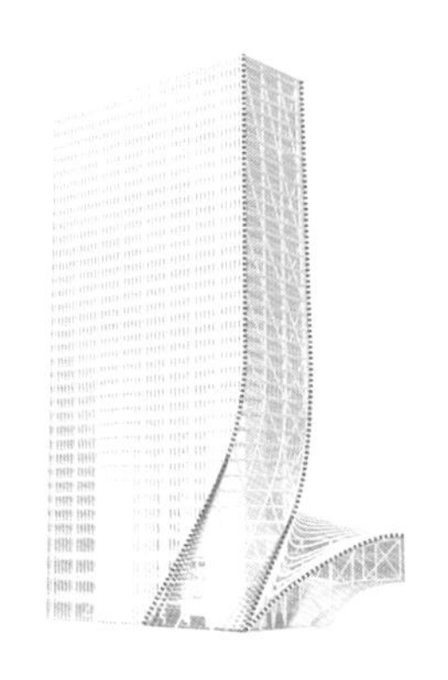

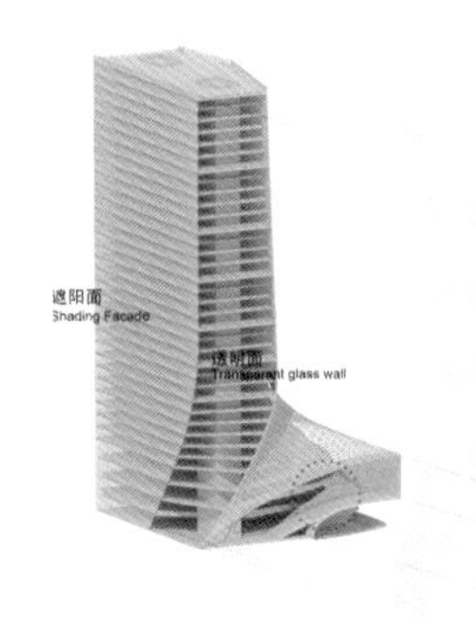
遮阳面
Shading Facade
透明面
Transparent glass wall

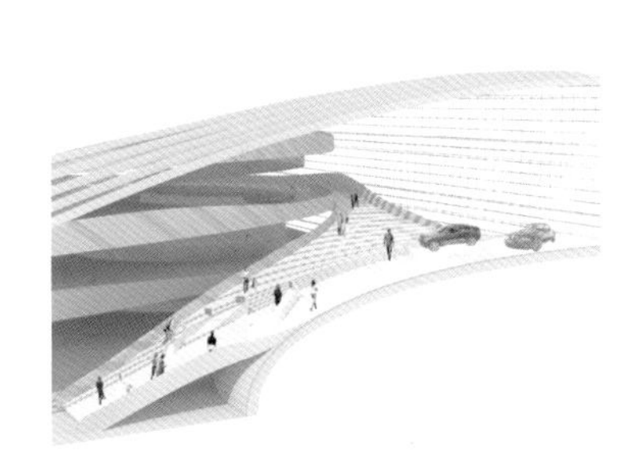

地铁前海湾车辆段上盖物业 深圳

Qianhai Bay Metro Depot Development, Shenzhen

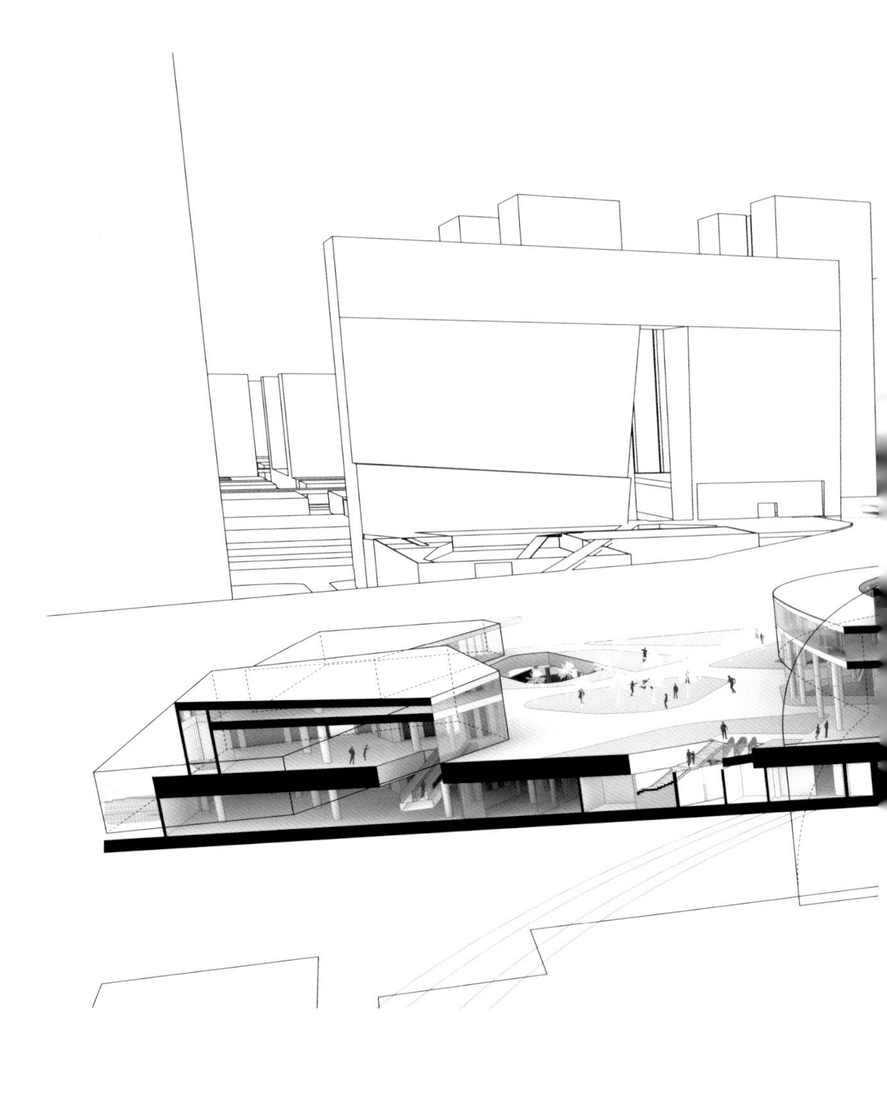

设计时间 / Design：2007 至今

用地面积 / Site Area：200000m²

建筑面积 / Building Area：354910m²

项目组 / Design Team：刘晓都 孟岩 | 徐罗以 | 林海滨 | 魏志姣 | 罗仁钦 张新峰 李昊 黄煦 艾芸 孙艳花 温谨馨 黄中汉 林恬 臧敏 席江 吴春英 邓军 何勇 张海君 涂江 饶恩辰 陈耀光 夏淼 左雷 王衍 傅卓恒 周娅琳 刘思钊 沈振中 李强 | 林挺 于晓兰 刘洁 陈丹平 廖志雄 胡伊硕

合作 / Collaborators：北京中外建建筑设计深圳分公司（施工图）

业主 / Client：深圳市地铁集团有限公司

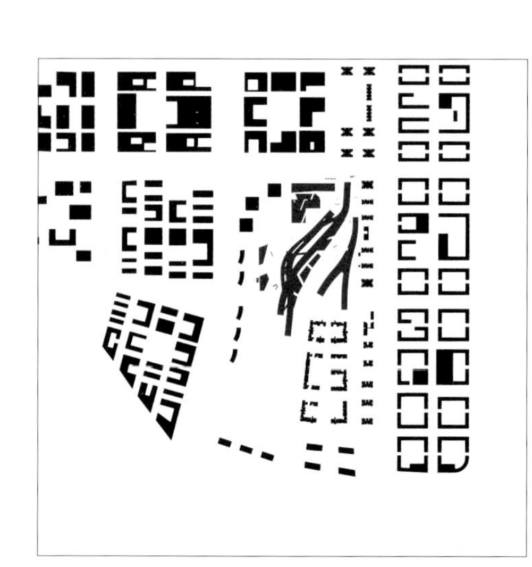

当深圳在城市规模上迅速跻身于国际大都市行列，其经济力量不断扩展和提升的时候，却日益显现出城市基础设施不足与用地极端缺乏的困境。这个矛盾在地铁建设上尤为突出。例如每条地铁线都有占地巨大的车辆段检修场，侵占了城市中心区宝贵的土地资源。为集约利用土地，深圳开始借鉴日本、中国香港等地区地铁上盖物业的做法，充分发挥地铁上盖的作用。前海上盖项目的如期建成，可能是国内首批此类工程，它将成为一种新的复合性城市多功能体的样板。

上盖物业的建设是基于已经设计完成的车辆段，只能在现有轨道之间见缝插针地设计结构。这样的限制，反而因势利导地促成了采用与下层轨道吻合的流线型来塑造出一个动感丰富的上层 SOHO 办公区，采用与限制条件高度统一的形式，恰如其分地表达下面看不到的真实。种种设计条件加大了设计的难度，例如荷载的限制使建筑高度有限而不能任意表现；不规则的柱网和超大的尺度使设计的性价比要认真考虑；当前建筑规范不适合于此类建筑造成的平面布置极其困难，等等。这些难度都使设计面临着空前挑战，而成功地应对这些挑战，恰恰展现了当代城市条件所能赋予当代建筑的魅力。

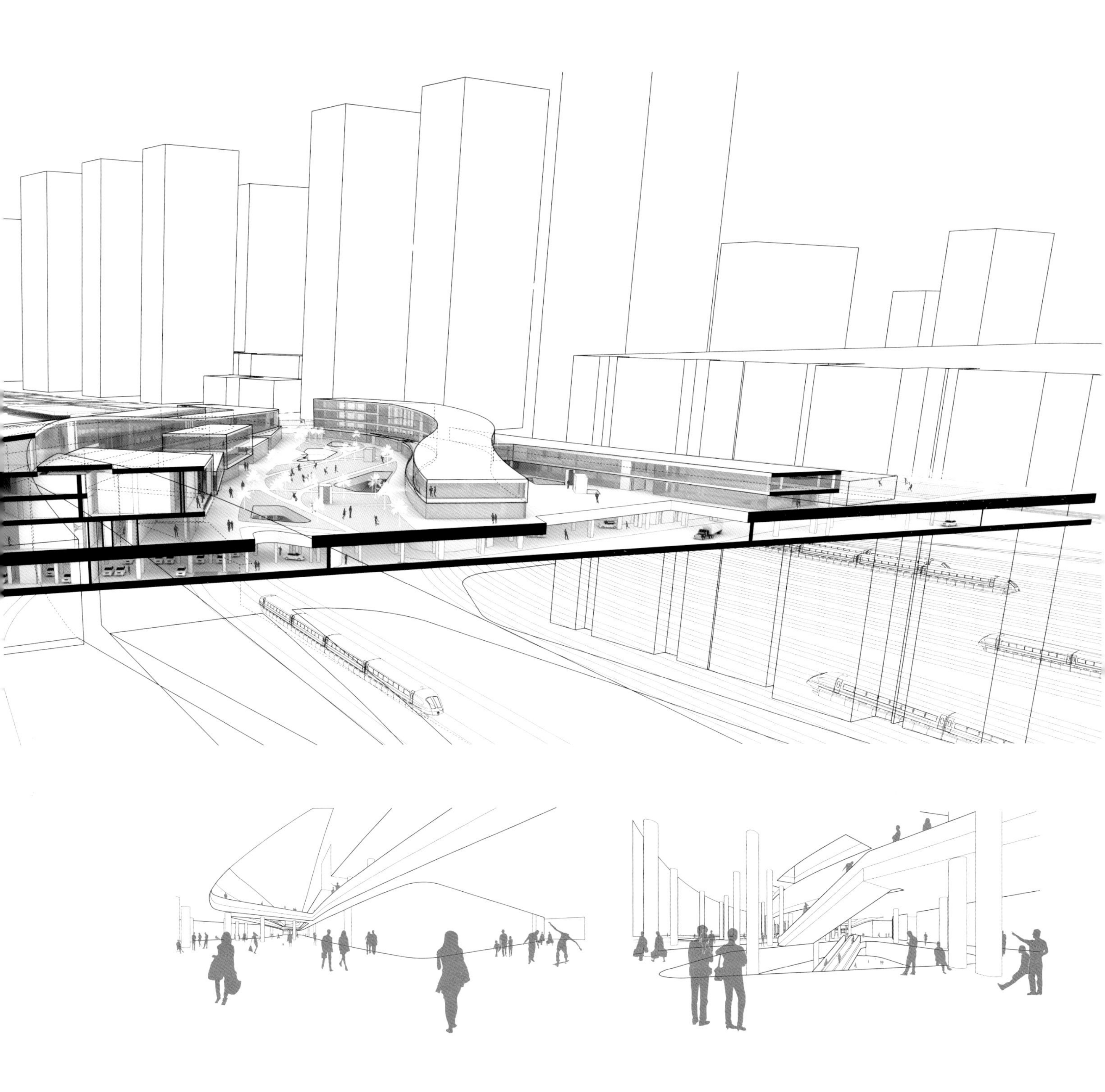

Shenzhen is rapidly becoming an international city with the help of its incessant economy and development, but the lack of city infrastructure and land for development has become an increasing problem. The challenge is clear when it comes to the construction of Shenzhen's Metro lines. Each line typically demands a large service depot in a valuable urban area, therefore making this area very inefficient for urban development. The city had decided to take on the idea from places like Japan and Hong Kong to develop a commercial complex on top of the metro service depot. After its completion, the Qianhai Complex on Top of Metro Service Yard will be the first of its kind in China, making it a template for new multi-functional urban complexes.

When work on the building-on-top was confirmed, the metro service depot had finished the design and was about to start construction. The challenge was to design the column network within the limited spaces between the rail lines, which actually brought an opportunity for us to incorporate the streamlined shape of the rail layout into our building design. It created a real vivid and dynamic form and place to fit the SOHO building type. The surface style echoed the function underneath perfectly. It was not easily achieved. Due to the load limit we were restricted to build only low-rise buildings. The fact that the ground level was raised at a height of 16 meters made it difficult to work out the circulation of the complex and to cooperate with the building codes. The irregular column grid and the huge site made the design process overwhelmingly difficult and time consuming. It has become the most complicated project that URBANUS has ever worked on.

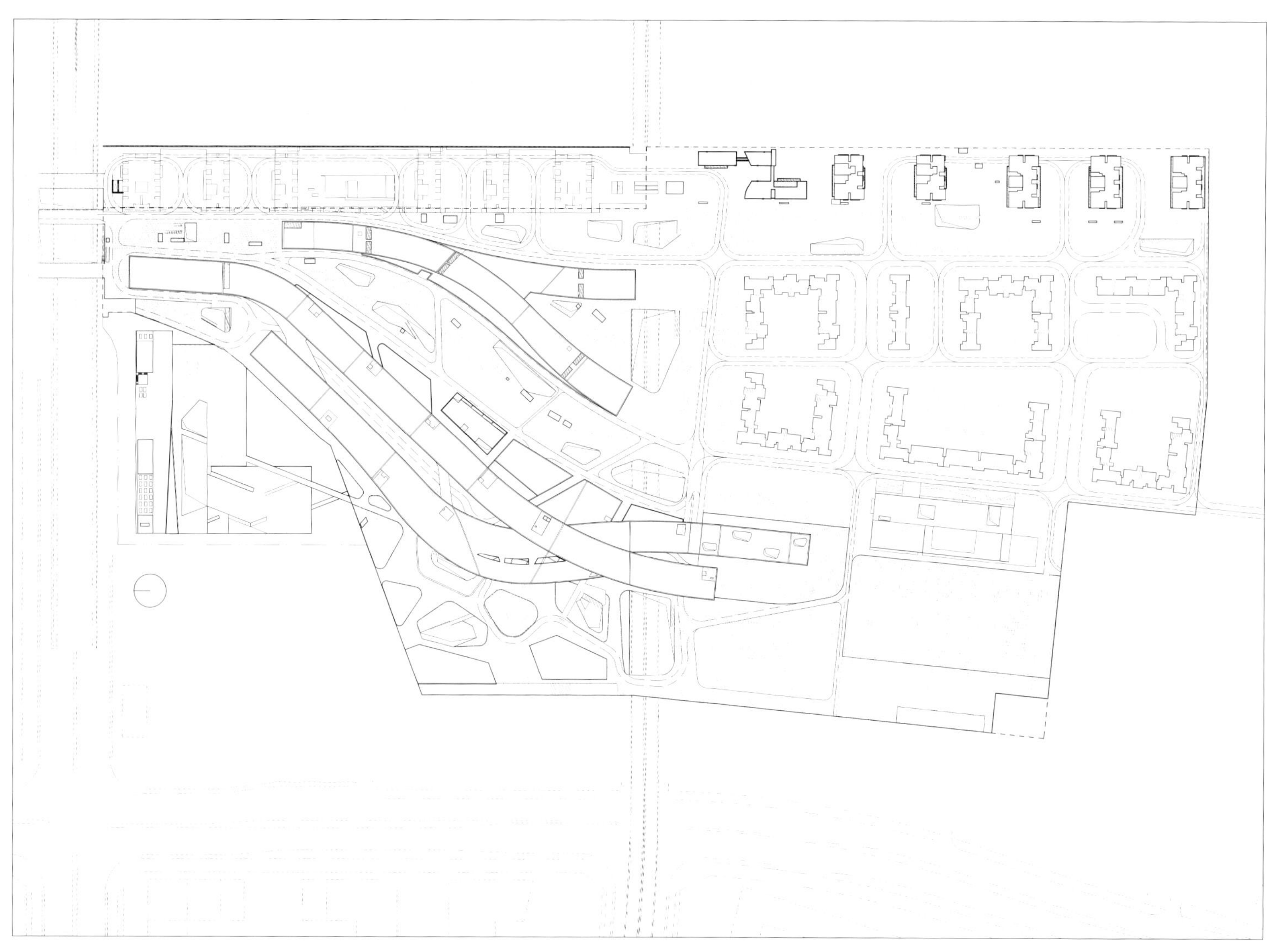

总平面图 / Site plan

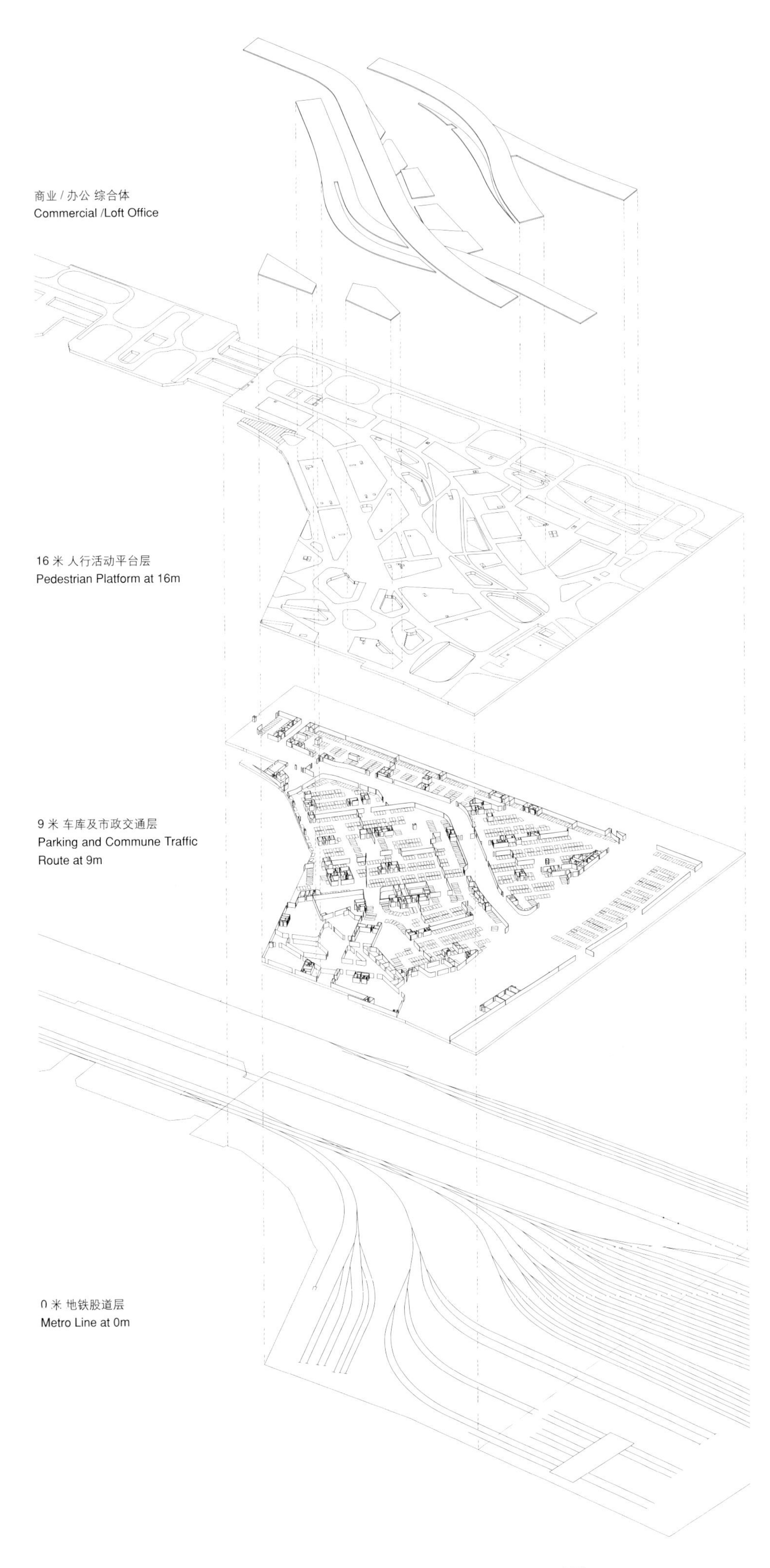
商业 / 办公 综合体
Commercial /Loft Office
16 米 人行活动平台层
Pedestrian Platform at 16m
9 米 车库及市政交通层
Parking and Commune Traffic
Route at 9m
0 米 地铁股道层
Metro Line at 0m

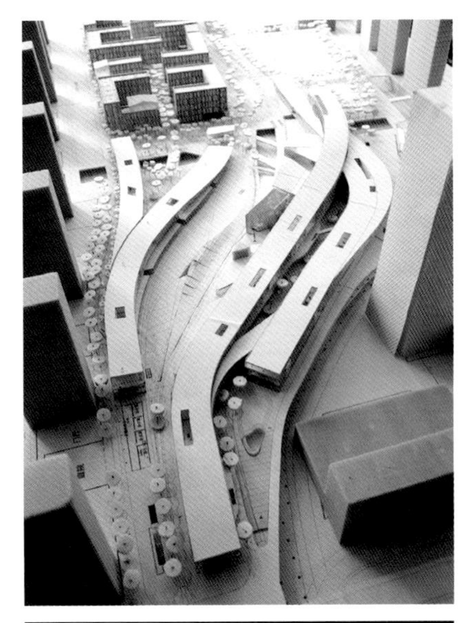

深圳眼（水晶岛）规划设计

Shenzhen Eye (Crystal Island) Urban Design

设计时间 / Design：2009 至今
用地面积 / Site Area：360000m^2
建筑面积 / Building Area：361000m^2
项目组 / Design Team：孟岩 刘晓都 | 苏爱迪 张天欣
梁广发 | 胡伊硕 张丽娜 罗仁钦 曾冠生 林秀清 成直
Samuel T. Ruberti
合作 / Collaborators：OMA 大都会建筑事务所
业主 / Client：深圳市规划和国土资源委员会

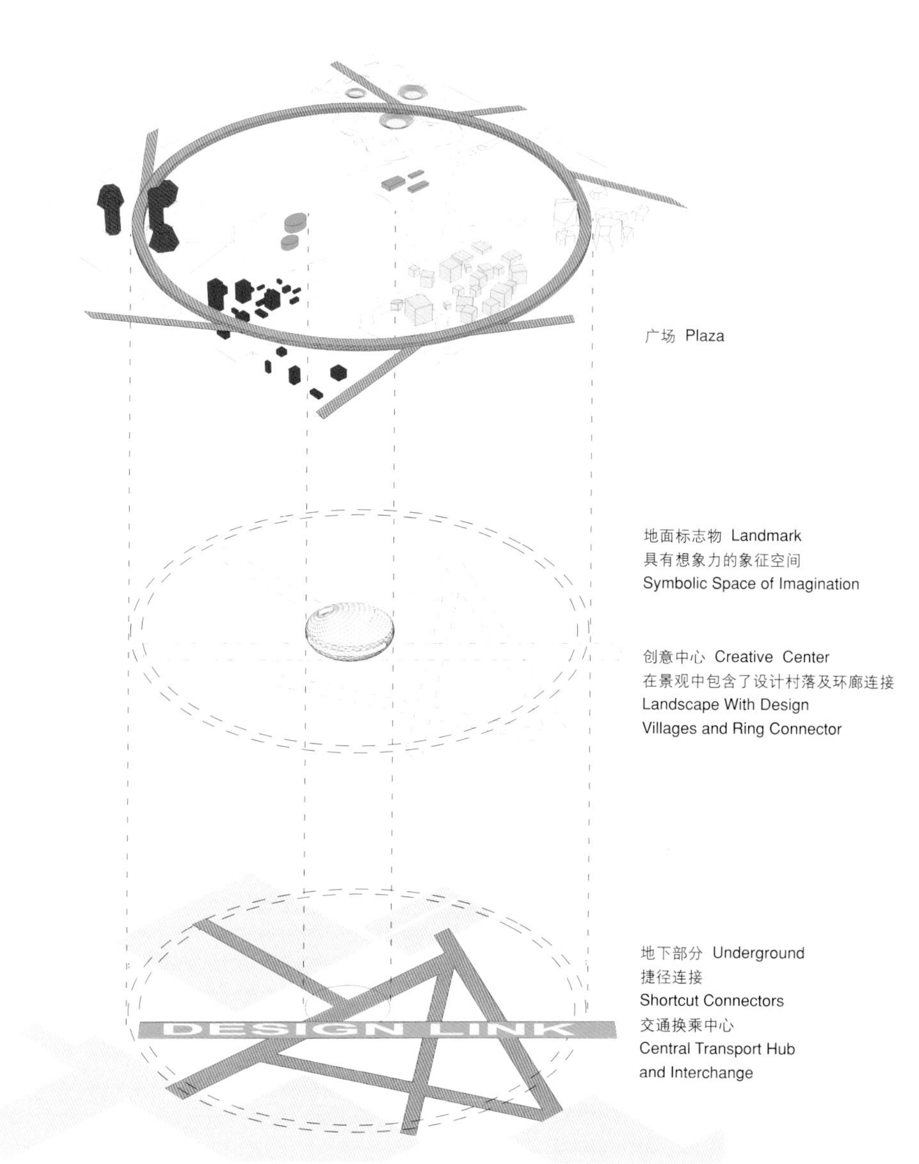

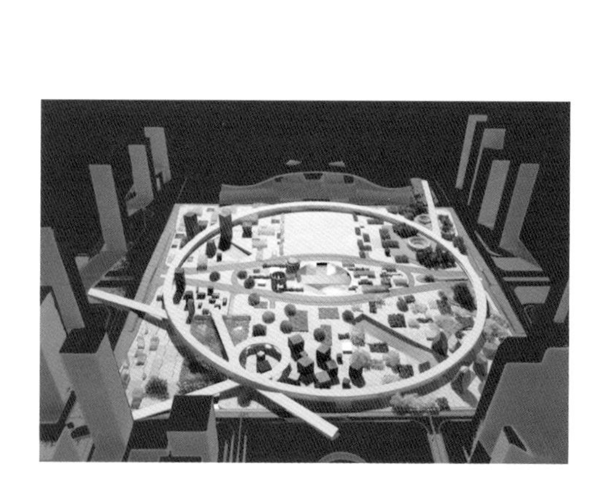

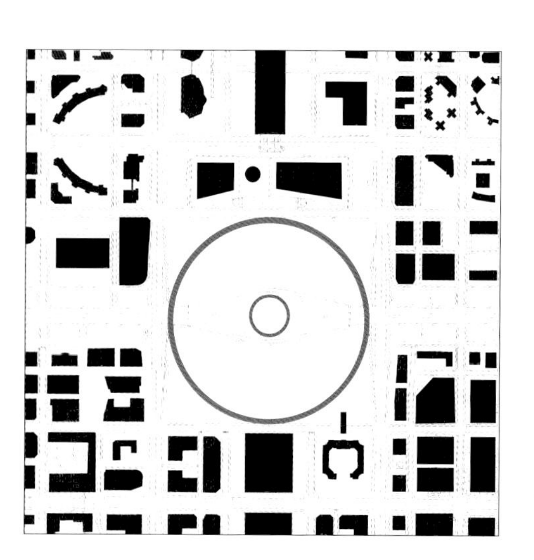

水晶岛位于深圳中心区的核心位置，是城市南北中轴线与东西主干道——深南大道交汇处，规划范围达 45 公顷。基地覆盖大片绿化，地下是施工中的未来深圳中央火车站的起点，及地铁一号线与巴士总站交汇处。项目的核心问题在于如何挖掘和利用其作为整个城市交通核心枢纽地带的价值和潜力，为城市未来的发展带来具有战略意义的规划设计。

深圳在过去的短短 30 年内从小渔村发展成一个经济中心和大都会城市，实现了第一次飞跃。当前所面临的飞跃是该如何跨越作为制造业中心的历史，为自身在全球化语境下重新定位。2008 年，深圳被联合国教科文组织授予“设计之都”的称号，城市创意资源丰富，但缺乏一个中心：一个能汇聚现有创意网络与能量的焦点。

这个设计的理念是将原本罕有人至的中心区空旷场地，转变成为设计之都的中心。它的地下空间由捷径系统构成，连接铁路和地铁车站，形成中央交通枢纽；地上部分架高的环形人行道连接体将建立有效的道路循环网络，连通作为创意景观与实体的设计村落和其他地上活动集群。这个网络的核心是一个属于心灵的空间，是自由的想象力和开放的创造力的精神所在。这个具寓意的“想象空间”，汇聚并激发深圳的能量，预示着深圳的将来，故而名之“深圳眼”。规划的最终目标在于为城市的创造力提供施展的机遇，为城市的未来提供发展的平台。

The Shenzhen eye project is located in the Shenzhen CBD, at the intersection of the north-south central axis and the Shennan Road, a major road connecting east and west. The planning site measures 45 hectares and is covered mainly by landscape and vegetation. The underground area is a vital site which is currently under construction for a future central traffic interchange hub, connecting the bus terminus and subway line 1 as well as the starting point of Shenzhen's central train station. The core aspect of the design is how to excavate and use the city's traffic hub efficiently, while at the same time providing strategic planning for the city's future development.

Shenzhen has progressed with astonishing speed in the past 30 years, transitioning from a fishing village into an economic center and international metropolis. How can Shenzhen transgress its history as a center for manufacturing and accelerate the development of its creative industries and find its new identity? In 2008, Shenzhen was awarded the title of "City of Design". However, even with rich creative resources and background, the people of Shenzhen lack a center, a focal point for the existing net-works and energies to converge, and a stage on which to communicate and inspire. Thus the construction of an underground system of Shortcut Connectors linking existing and future train/subway stations to create a Central Transport Hub and a Creative Landscape with Design Villages and clusters of activity above-ground that are joined by an elevated Ring Connector.

The center of the interwoven elements is a mind space: free, open, and unoccupied. It is a void of imaginary power, a spiritual space of creativity that bundles and concentrates Shenzhen's energy and vision for the future: Shenzhen Eye. The combination of these elements in the Shenzhen Creative Center captures, focuses and redistributes the diverse energies of the city, and provides a stage for encounters, activities, and the further growth of its creative forces.

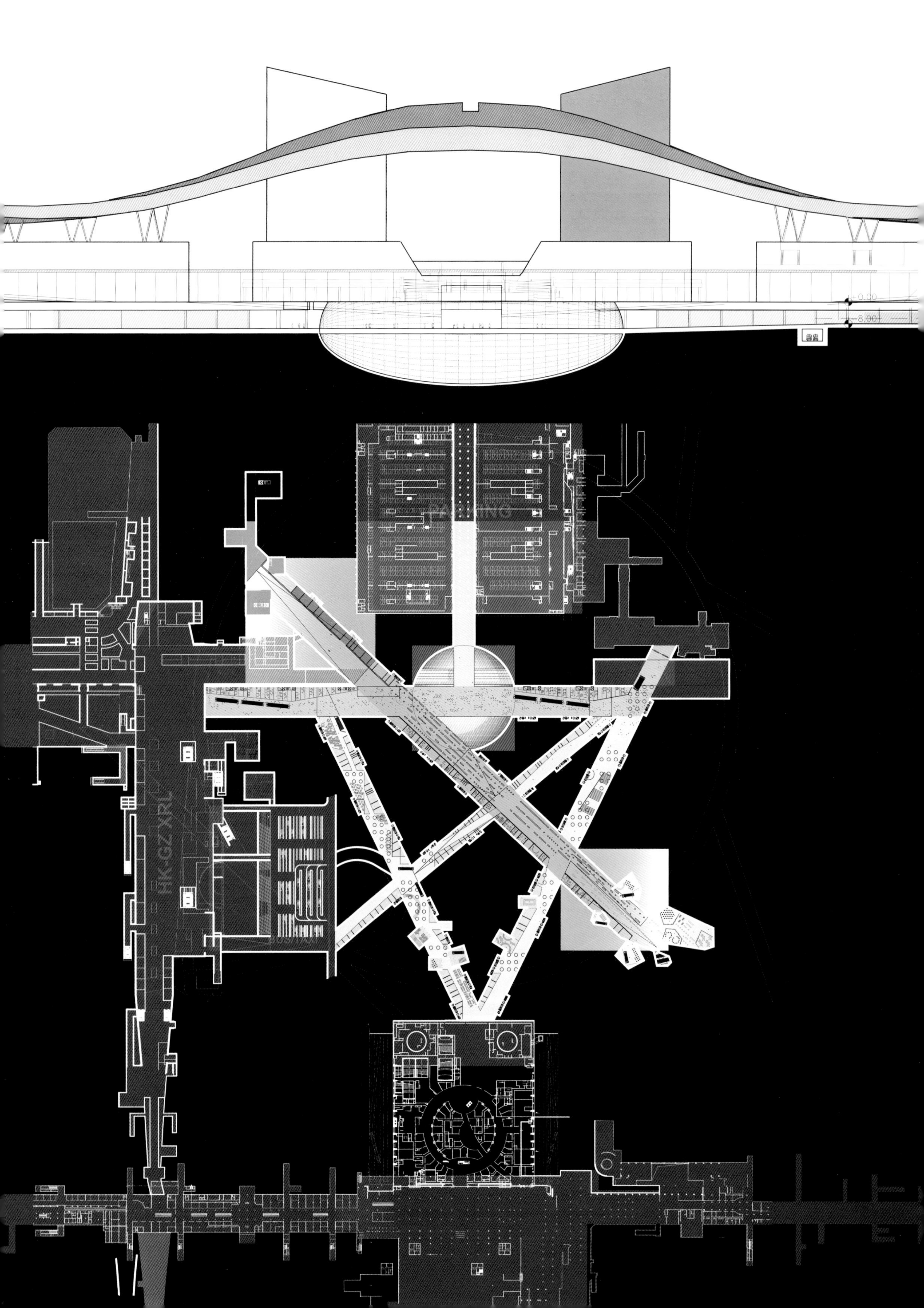
+0.00
-8.00
PARKING
HK-GZ XRL
BUS/TAXI

华强北商业街区更新计划 深圳

Huaqiangbei Commercial Area Renewal Plan, Shenzhen

设计时间 / Design：2009
用地面积 / Site Area：45hm²
项目组 / Design Team：孟岩 刘晓都 | 王衍 张丽娜 何勇 梁广发 欧阳祎 黄艺宏 熊慧晶 魏志姣
合作 / Collaborators：深圳市城市规划设计研究院
业主 / Client：深圳市福田区人民政府　深圳市规划局

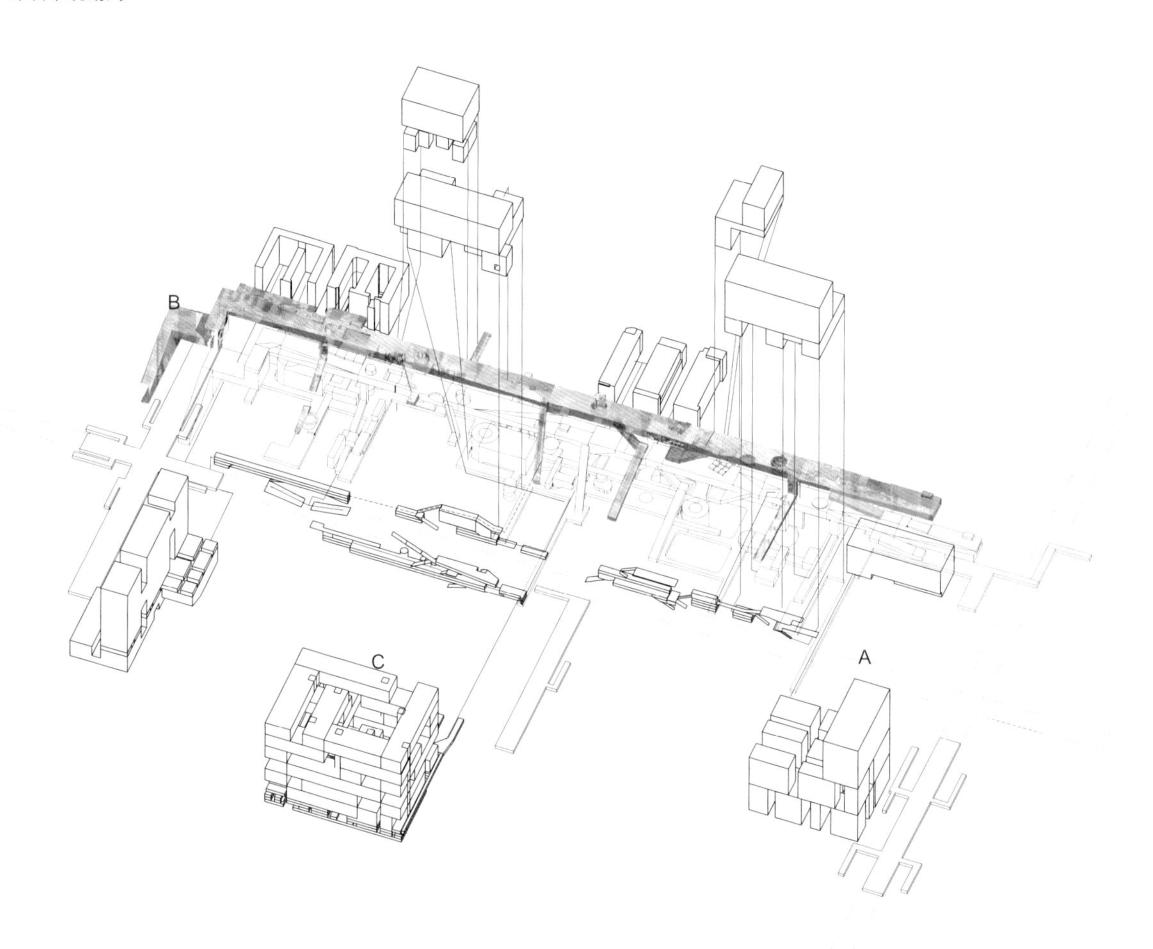

A：华强北
Huaqiangbei Road
B：基础·街
Infrastreet as a Linkage
C：嘉年世纪广场
Jianian Century Plaza

作为深圳主要核心商圈之一，华强北从一个电子工业制造基地，转变成聚集IT产业链和各种商业配套服务的著名电子商业街。工业厂房街区的特殊空间肌理，使种类繁多的零售商业、物流系统及各种配套服务业得以向街区纵深延展，造就了华强北的繁荣。近十年来，大型商业综合体入侵原厂房街区肌理的趋势不断加剧，封闭内向、模式单一的购物中心模式与原本生长在内部街区连续、多样的小商业模式之间的矛盾日趋激烈，原有的街区活力正在面临着丧失的危险。未来几年，4条地铁线将在华强北商业区地下贯通，长度仅为940米的华强北路将密集出现3个地铁站点。如何在新一轮的大开发中寻找到一种新的、混合的城市更新发展模式，在看似密集的大开发下保护华强北商业圈独特的空间特质和包容性，为小尺度的商业提供容身之所？

此城市设计的核心问题表面是缓解华强北交通压力，其实是梳理街区内部的以及两侧街区之间的合理的立体联系。由此，我们提出"基础街"作为基本策略，希望在侧街建造一个连续的提供辅助公共功能的混合"基础设施"，包括连桥及整体地下商业街的开发，为两侧沿街商业提供有效的辅助，使其能够容纳原本活跃的小商业。新的开发由"基础街"连接，成为系统中的活力点。"基础街"的策略是一种引导在华强北路两侧进行高密度商业开发的立体城市空间策略，旨在为华强北商业圈提供合理而独特的城市空间更新模式。

The Huaqiangbei district is one of the major commercial areas in Shenzhen. In the last decade, it transformed from an electronics manufacturing industrial cluster into a famous trading hub for electronics connecting to many industrial networks and supported by comprehensive business services. Existing industrial factories in this area provide a unique spatial condition for the emergence and development of various retail traders, as well as logistics industries and their related businesses. Recently, more large-scale shopping malls have occupied blocks of Huaqiangbei's industrial urban fabrics. The urban blocks became introverted and single programmed.

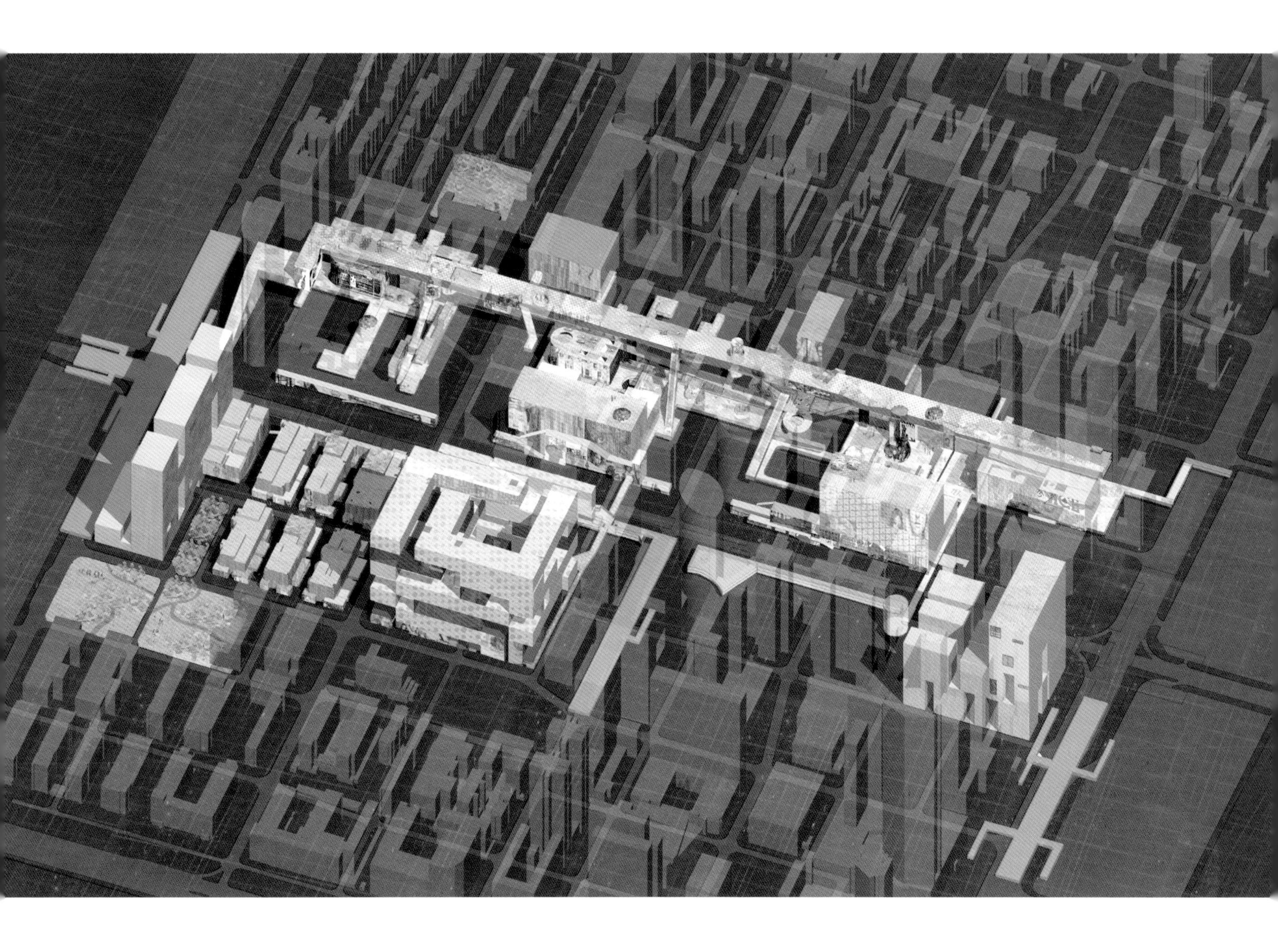

This mutation contrasts to and conflicts with the previous industrially built environment, which is recognized by porous, multi-programmed back street spaces with many small-scale retailers. Consequently, the previous active urban space is declining. Very soon, four metro lines will pass through the site, and three metro stations will be built on the 940 meter long Huaqiangbei Road. This planning indicates that inevitably another wave of large scale development will happen in the Huaqiangbei commercial district. Knowing that Huaqiangbei is facing this developmental challenge, it is urgent to find a new way of spatial organization and design that emphases mixed use.

In front of the background of an inevitable massive development, we must question how to preserve and encourage the inclusive and human scale urban space, but also to support the large-scale development of the city.

On the ground level, the problem in Huaqiangbei is the inextricable traffic flow. Based on our research, the solution of such a problem is to regulate the intra-block organization and to redefine a reasonable 3-dimensional inter-block connection for pedestrian. Therefore, we introduce our strategy of INFRASTREET. As our main goal, we intend to build public service infrastructure to connect and penetrate the blocks. It will serve Huaqiangbei Road, facilitate internal circulations of the blocks, and mediate existing and potential problems of high-density development. Large scale developments should be properly guided; as urban anchors they need to have the capability of contain the small-scale business. At the same time, the INFRASTREET system will connect big business anchors on both sides of the road. To conclude, INFASTREET is a spatial strategy encouraging appropriate large scale development and is the seed for suitable and proper potential development with a human scale.

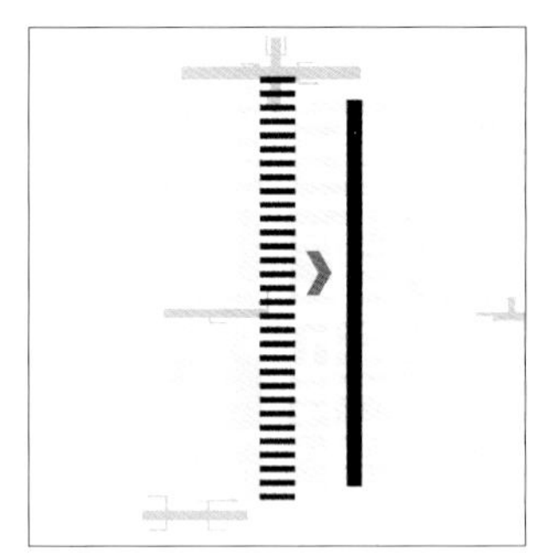

Phase 1

大开发应得到正确的空间引导，由“基础街”连接，成为系统中的活力点。

High-density mass development should be guided properly. As urban anchors, they are connected by the INFRASTREET.

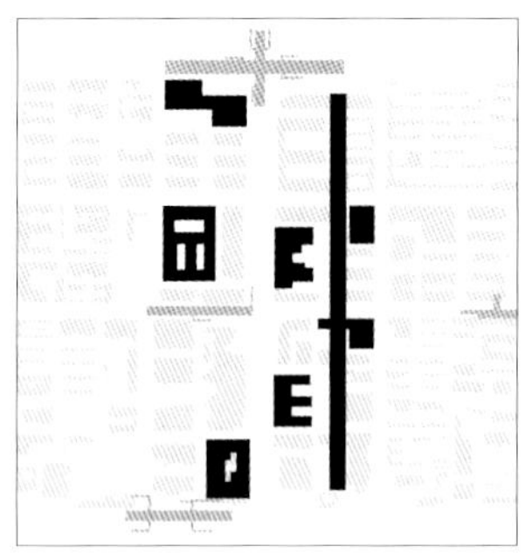

Phase 2

建造一个提供辅助公共功能的构筑物——“基础街”连接东侧的街区，作为基础设施，为未来两侧沿街商业面提供有效的支持，将改善华强北的现存问题，支持未来的大开发。

We intend to build an INFRASTEET working as a public service infrastructure in the block. It serves Huaqiangbei Road, helping circulation in the blocks, mediating existing and potential problems in terms of mass development.

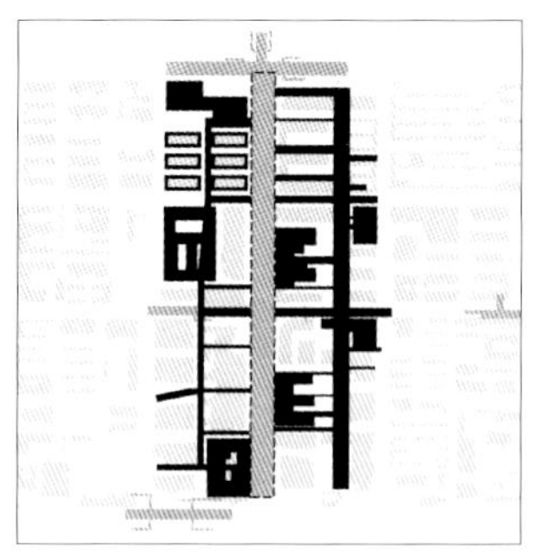

Phase 3

华强北路上的立体街道包括了连桥和整体地下商业街的开发，这将成为连接两侧街区的有效支持。

Development including bridges and integrated underground shopping street will connect anchors in both side of the road.

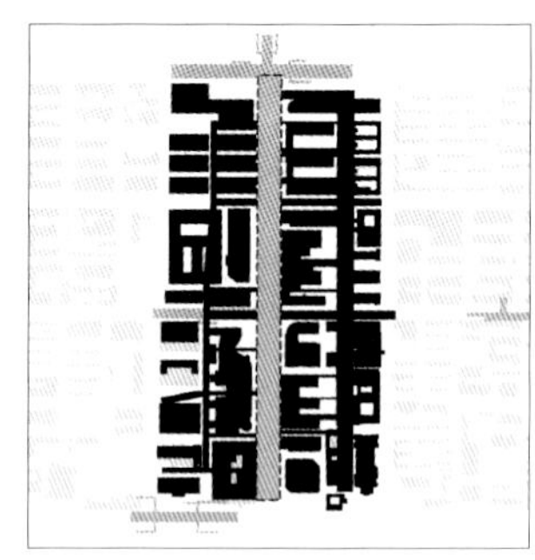

Phase 4

这是一种鼓励华强北路两侧高密度开发的立体城市空间策略，引导这些开发，为华强北商业圈提供适合并合理的城市空间模式。

This is a spatial strategy encouraging appropriate high-density mass development for the high-end urban space at both side of the Huaqiangbei Road. It is a space manual for these potential developments.

C：嘉年世纪广场
Jianian Century Plaza

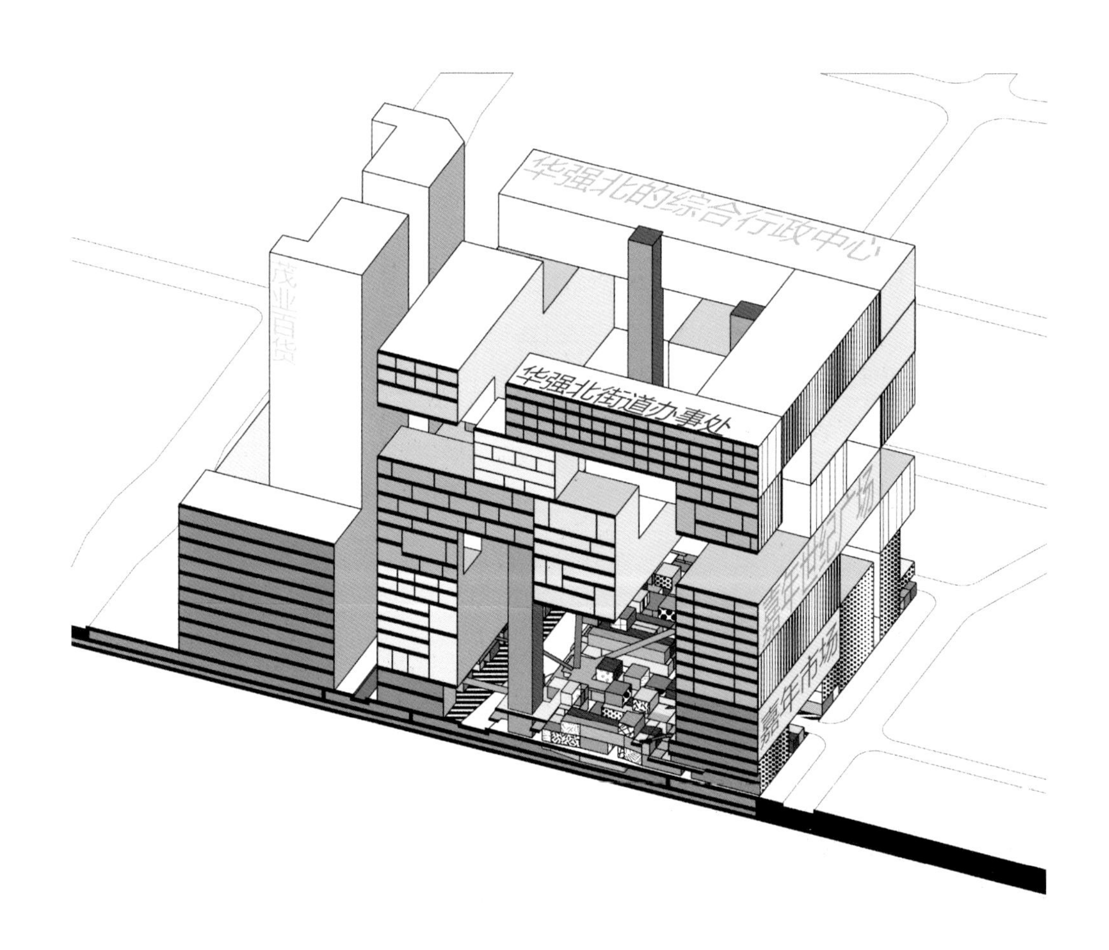

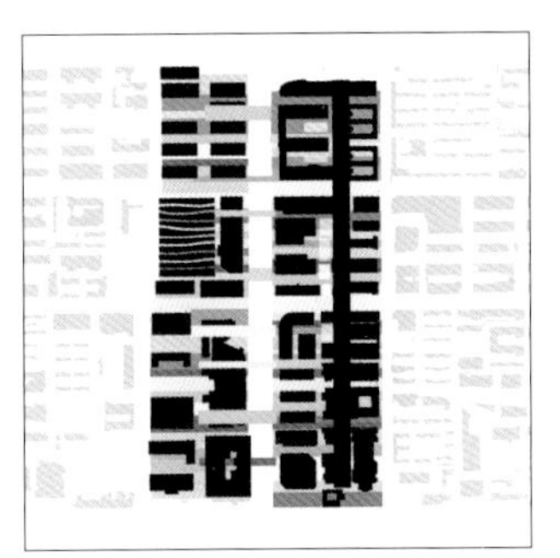

Top-down and Bottom-up strategy

在看似密集的大开发下保护华强北商业圈现有城市街区内独特的空间包容性。

Under the seemingly mass develepement, the uniqueness of commercially positive streets low-end context which identifies urban form of old Huaqiangbei industrial zone is preserved.

见缝插针的小公园成为步行系统中最具活力的公共空间
Those little slot parks are the most attractive public space in the 3d walking system.

人与小型公共空间的置入
Human Scale. Slot Space

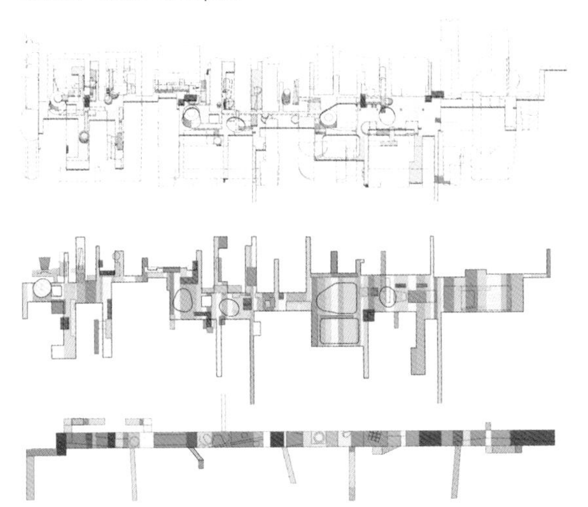

- 百货 / 零售 / 跳蚤市场 / 电子交易中心餐厅 / 咖啡吧 / 小吃街 / 宿舍
- 住宅会议中心 / 图书馆 / 电影院 / 剧场 / 画廊 / 小会堂
- 篮球场 / 太极 / 私家花园 / 院落 / 太阳浴 / 健身房
- 公司总部 / 办公 / 仓库
- 艺术家工作室 / LOFT / 雕塑公园
- 公共空间 / 城市广场 / 休息室

嘉年世纪广场要同时保证大开发和嘉华小市场的生存
Jianian Century Plaza can allow the coexistence between big development and low-end market

张江高新技术园区嘉定园城市设计 上海

Urban Design for Zhangjiang High-Tech Park Jiading Branch, Shanghai

设计时间 / Design：2011

用地面积 / Site Area：100000m²

项目组 / Team：刘晓都 | 林达 | Danil Nagy Julianna Kei Michael H. Rogers Cressica J. Brazier

业主 / Client：上海市嘉定区菊园新区管理委员会

转变中的研发空间研究
Transforming R&D Workspace

1960-
传统综合体
Traditional Complex

1970-
租赁办公
Office Rental

1990-
向小型化发展
Small Startups

2000-
远程办公
Telecommuting

切入点
Incenception

联合工作共享办公
Co-Working Communal Office

高科园空间的多元模式研究
Multiple Patterns of R&D Campus Space

1960-
传统园区
Traditional Campus

1970-
科技园社区
Campus Community

1990-
公共科技园
Public Tech Park

多样规模园
Multi-Scale Park

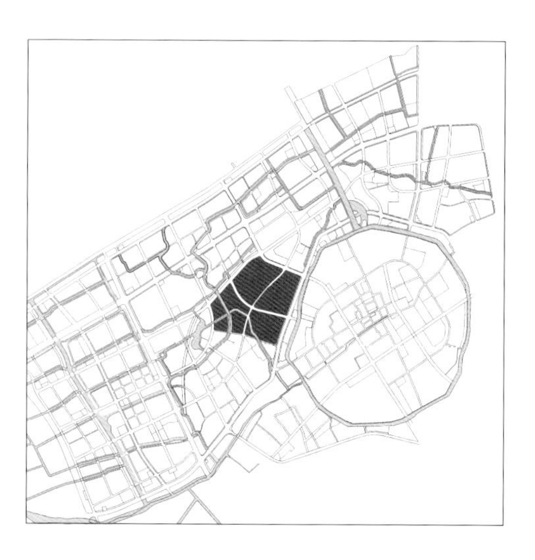

上海张江高新科技园（嘉定园）位于上海北部的嘉定区。设计反思中国科技产业园的模式及类型，基于基地上现有产业活动，对其进行升级并创造更富创意的新型科技知识枢纽。

传统科技园区的设计往往导致大规模、相互分隔及单一用途的园区，不能鼓励科技产业之间的互动，更与城市系统脱节。此设计除了积极打造研究中心外，还为周边社区提供公共空间及社区功能。本方案的结果基于我们对中国未来产业发展及工作空间的研究。我们发现，在这十年间，由急速增长的出口产品制造产业带动了中国城市进行大规模的产业结构转型。为了在未来创造更具可持续性的经济发展，中国本土产业也逐渐由净生产模式向产品研发及生产的混合模式进化。

为了促进更多的协作关系，我们的城市设计提倡一种由混合功能、多元尺度建筑构成的新型产业园区模式。其一，在园区基地中有十块土地为科研机构的入驻提供场地。这些土地可以满足研发活动中比较私密的要求。其二，更为公共的功能组织在园区的中心，形成商业核心。这些公共功能包括会议空间、活动空间及短期居住空间等，供园区内的成员分享。其三，利用由中心商业区与城市之间的河道及廊道等八条各具特色的“触须”令园区与城市紧密联系。最后，我们提倡改造园区中的旧建筑，成为小型企业孵化器功能的板块。这个孵化器为园区中大型科研机构的上下游分支企业提供基地，孕育未来的成功企业。

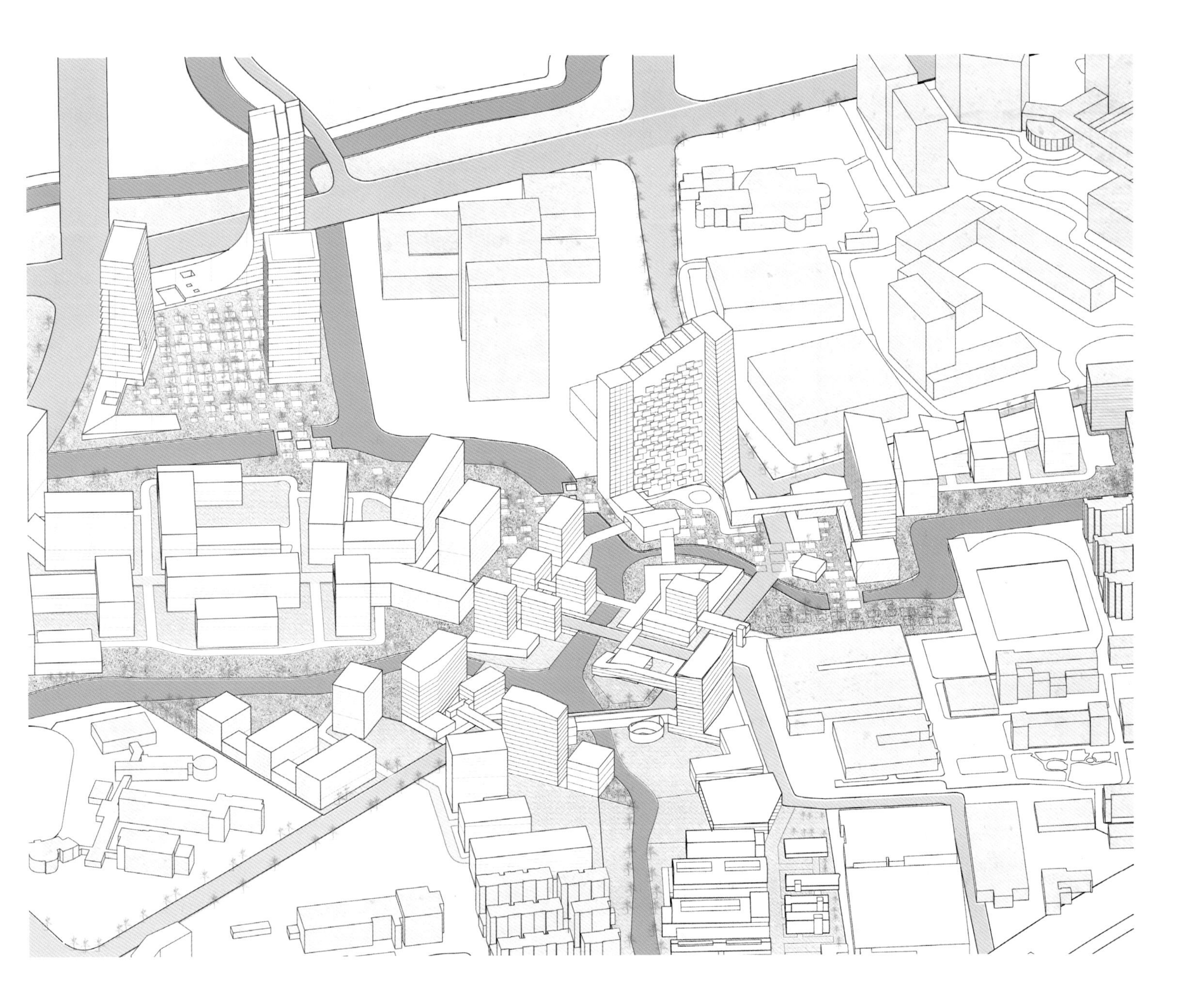

Zhangjiang Technology Park reinvents the typology of a science and technology park development in China. Located in the Jiading district north of Shanghai, the park capitalizes on the area's existing base of industrial development to create an innovative model for a new era of technology hubs.The project responds directly to the traditional form of technology park planning, which has resulted in large-scale, isolated, single-use campuses that fail to create a collaborative research environment and do not connect to a supportive urban community. In response, the Zhangjiang Technology Park is designed to maintain its identity as a center of research while remaining open to the surrounding community through shared amenities and open spaces. The project also responds directly to research conducted on the future of China's industrial development and the types of work spaces necessary to support this development. Over the past few decades, China has been undergoing a massive industrial transformation, led primarily by rapidly increasing industrial capacity geared toward the production of goods for foreign firms and exported to foreign markets. In order to foster more sustainable economic growth into the future, the country is now shifting its focus toward fostering domestic firms with a strong emphasis on research and development.To promote collaboration, the Zhangjiang design proposes a new model of a mixed use, multi-scale innovation park consisting of three main parts. Ten smaller scale private parcels house research institutes and their more private R&D and administrative programs. More public support functions such as conference centers, meeting rooms, and accommodation are shared among park members and combined into a commercial hub at the center of the site. The hub connects to the site's circulation network, forming a series of unique "tentacles" which create an identity for the campus while also maintaining connections to the surrounding context. Finally, a business incubator is established within a rehabilitated industrial area of the site. This incubator provides work spaces for smaller spin-off enterprises as well as support services for incubator members, members of the park, and the larger industrial community.

策略 / Strategy

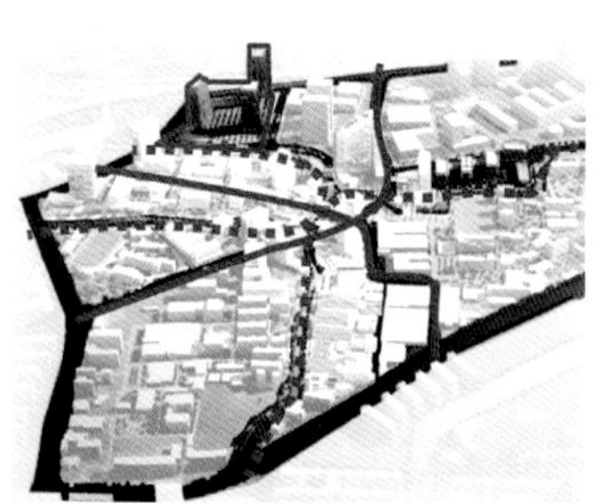

Phase I

办公楼组群
Office Cluster

发展道路系统和改造河道。基地上现有道路旁的地块可先被发展。
Develop road and canal systems. Parcels that is ad-jacent to existing roads can be developed in this phase.

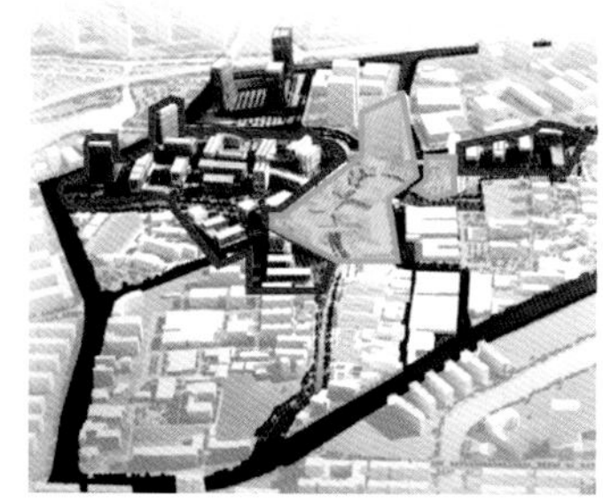

Phase II

研发地块
R&D Parcel

各地块可由不同的研究院各自发展，中心地块会为暂时的休闲空间和服务商业。
Parcels will be sold to different R&D institutes. The central area will become temporary green spaces and temporary retail business.

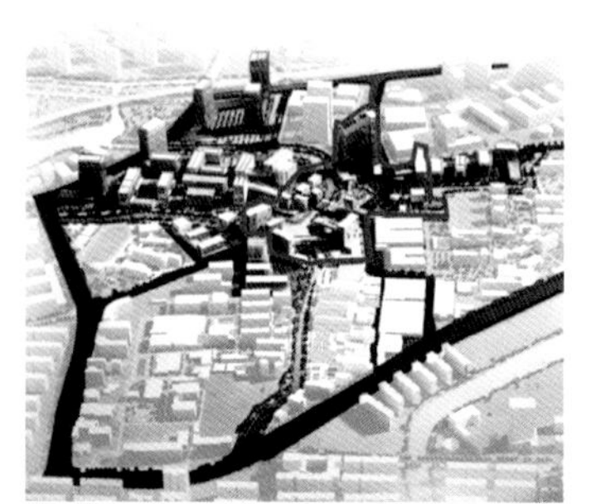

Phase III

中心枢纽
Central Hub

中心地块整体发展，包括住宿、会议中心、图书馆、零售庭院和孵化器。
The central parcel will be one development, including accomodation, convention, library, retai courtyards and incubator.

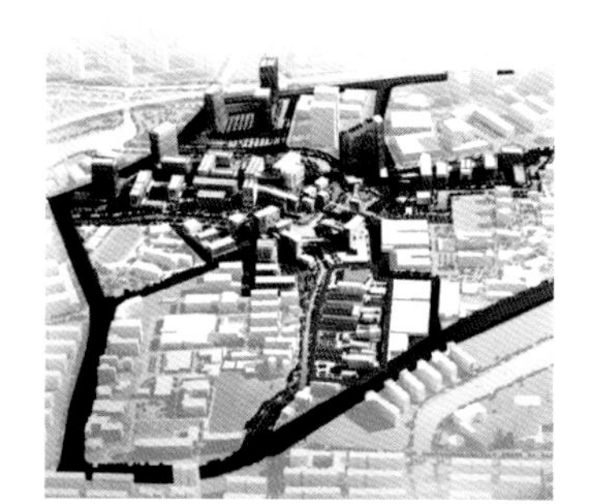

Phase IV

旧有工厂改造孵化器
Refurbished Factoryfor Incubator

取得基地上现有小型工商业用地，改造为高科技孵化器的建筑。
Acquired existing small scale industrial building on site and converted into Hi-Tech incubator buildings.

中心圈和枢纽 / Loop & Hub

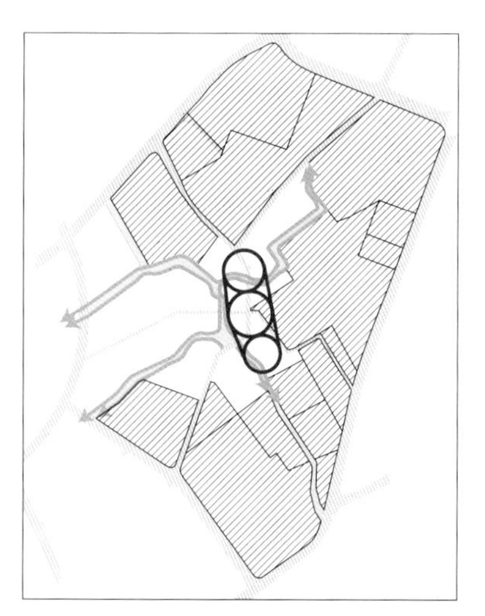

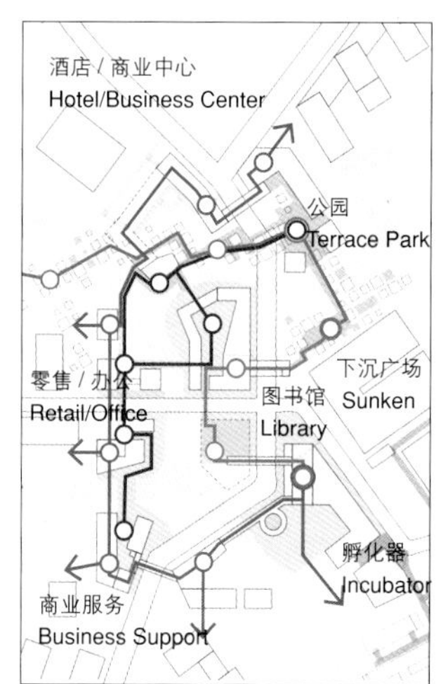

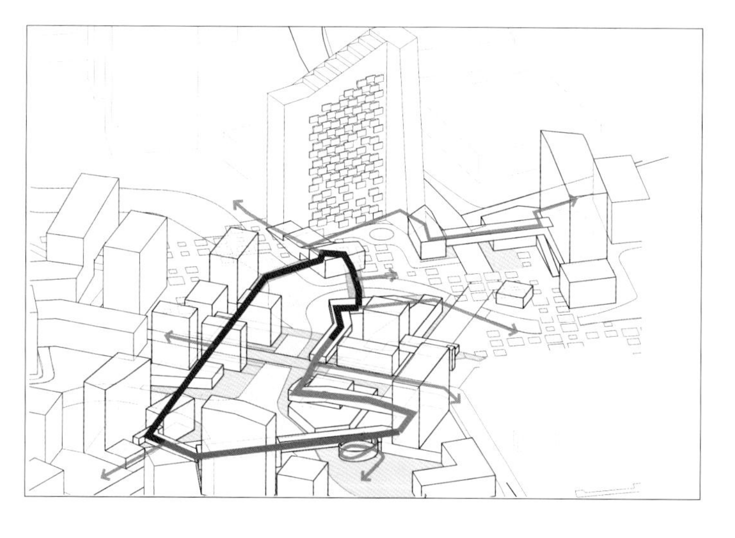

为了使复杂道路和运河在基地中心有很好的交接，枢纽发展成一个概念性的循环。这个循环可以接触围绕中心的各种重要的交接点和连接各基地的主导轴。

A central hub will negotiate the complex interchange of transit modes occuring. In order to negotiate the complex intersection at the center of the site, the idea of the hub is expanded into a conceptual loop around the center of the site. This loop can touch various important interchange points around the center and connect the dominant axes of the site.

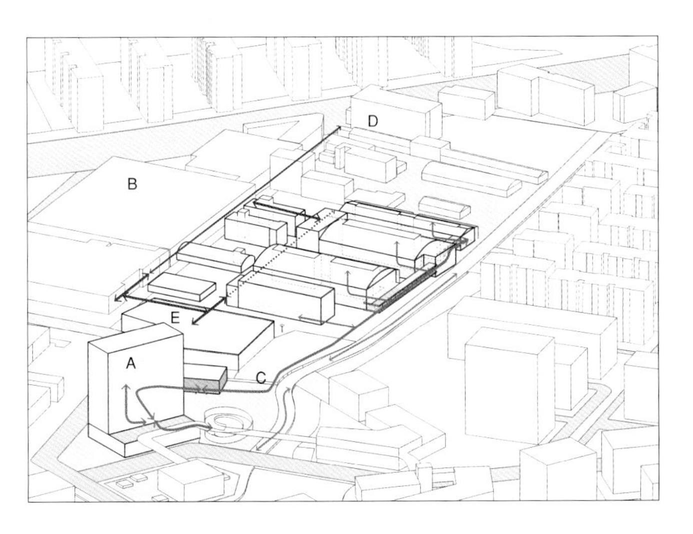

A: 孵化器大楼 Incubator Building
B: 服务及现有娱乐业 Services and Entertainment
C: 桥上画廊 / 展示厅 Ramped Gallery
D: 旧有工厂改造孵化器 Refurbished Factory for Incubator
E: 停车场 Garadge

中国国学中心 北京

China National Sinology Center, Beijing

设计时间 / Design：2011

用地面积 / Site Area：16200m²

建筑面积 / Building Area：70000m²

地上 50000m² 地下 20000m²

项目组 / Design Team：王辉 孟岩 | 吴文一 | 郝钢 黄志毅 | 饶恩辰 王俊 苏爱迪 张天欣 黄中汉 朱伶俐 赵洪言 李淳 林秀清 孔祥磊 Jennifer H. Ha Deborah A. Richards 刘爽 郑明璐

业主 / Client：中华人民共和国国务院参事室

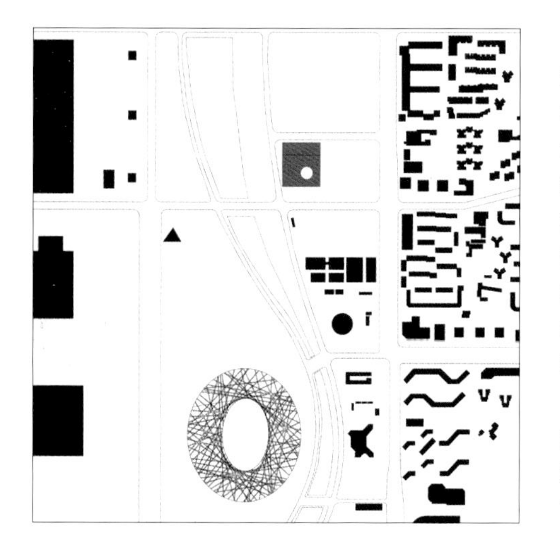

中国国学中心是国家最重要的文化机构，既是向世界展示和交流中华文化的窗口，又是对国人进行中华文化灌输和启蒙的场所。这一命题考量着设计者的立场与思想。我们着力于两点去研究问题的要害：一，当代国学的困境；二，当今表现中国精神的困境。当前，对国学的回归往往流于事件化、戏剧化、商业化，使国学成为一种远离大众日常生活的古老传统和学问。复兴国学，就应让国学重新回到大众的日常生活体验中。中国国学中心选址于国家奥林匹克中心这个都市场所，是实现这个目标的契机。

本设计理念是将国学理解为一个大众化的"都市国学"，把国学中心设计成为一个能够将传统国学对接到现实生活平台上的"国学都市"：在开放的首层与地下空间将国学中心丰富的展示、体验、教学内容与市井生活交融，并在立体画卷之中引导人流逐级上升，从繁华都市场景过渡到纯然静思的个人体验与精神洗礼。这些场景凝结在一个逻辑清晰的建筑空间构架中：其留白之处蕴藏了九曲盘桓的中国式空间；其间架结构的搭接，又有中国式建构的神韵。这个骨架之外，是层如玉的半透明玻璃外罩，半透出建筑内部叠构的体量和光影效果，并与多媒体影像叠合，构成一幅历史与现实交错、物质性与虚拟性互动的立面。这个外柔内刚的形体应和了"君子比德于玉"的传统文化心理。这种从精神的形而上的层面出发的造型处理，没有直接引用中国传统建筑符号，使国学中心不只是一个有中国元素的现代建筑，而且是一个有中国气度的当代经典建筑。

The National Sinology Center acts not only as an academic institute, but more importantly as a display window communicating the originality of Guoxue(Sinology), with a critical mission of preserving and spreading knowledge. The center plays the role of the most important ritual device, which challenges the design to analyze disposition possibilities and the implication of the abstract oriental spirit.The concept has raised two crucial problems in regards to the design. Firstly, research has determined that "Guoxue" has encountered a stage facing a critical crisis, genetically turns out to be commercialized. Secondly, it has been problematic to express and evaluate the traditional Chinese spirit while the manifestation of such topics has turned into incomprehensive fantasies that are most of the time remote from everyday life. The project, which is located in the national Olympic center, has provided opportunities to solve these issues.The design concept is to translate the inner spirit of "Guoxue" into an urbanized background. The open floor plan at the ground floor has mingled the generic urban lifestyle with exhibition and educational spaces, leading the flow of pedestrian traffic upwards which emphasizes the physical and metaphorical transition between chaotic urban scenes to sublimation personal experience.The scenes along the journey are carefully constructed via clear logical architectural spaces: the void spaces have a frame and tanon structure elaborating the "blankness" and "meditative" spirit lingering in traditional Chinese architecture. Enveloping the frame structure there is a façade made out of half translucent glazing imitating the luminous effect of jade. Layered with multi-media technology, the façade is an ambiguous display frontier showing a merging motion picture of historical archives and reality. The contrast between the softness of the façade and the solidness of the internal structure is amplifying the core Chinese cultural spirit, which is the inner peace as a "junzi"(gentalmen). The project has not deliberately imitated traditional patterns decorated with physical repetitive architectural symbols, but has manifested to revitalize the spiritual essence of Chinese cultural studies.

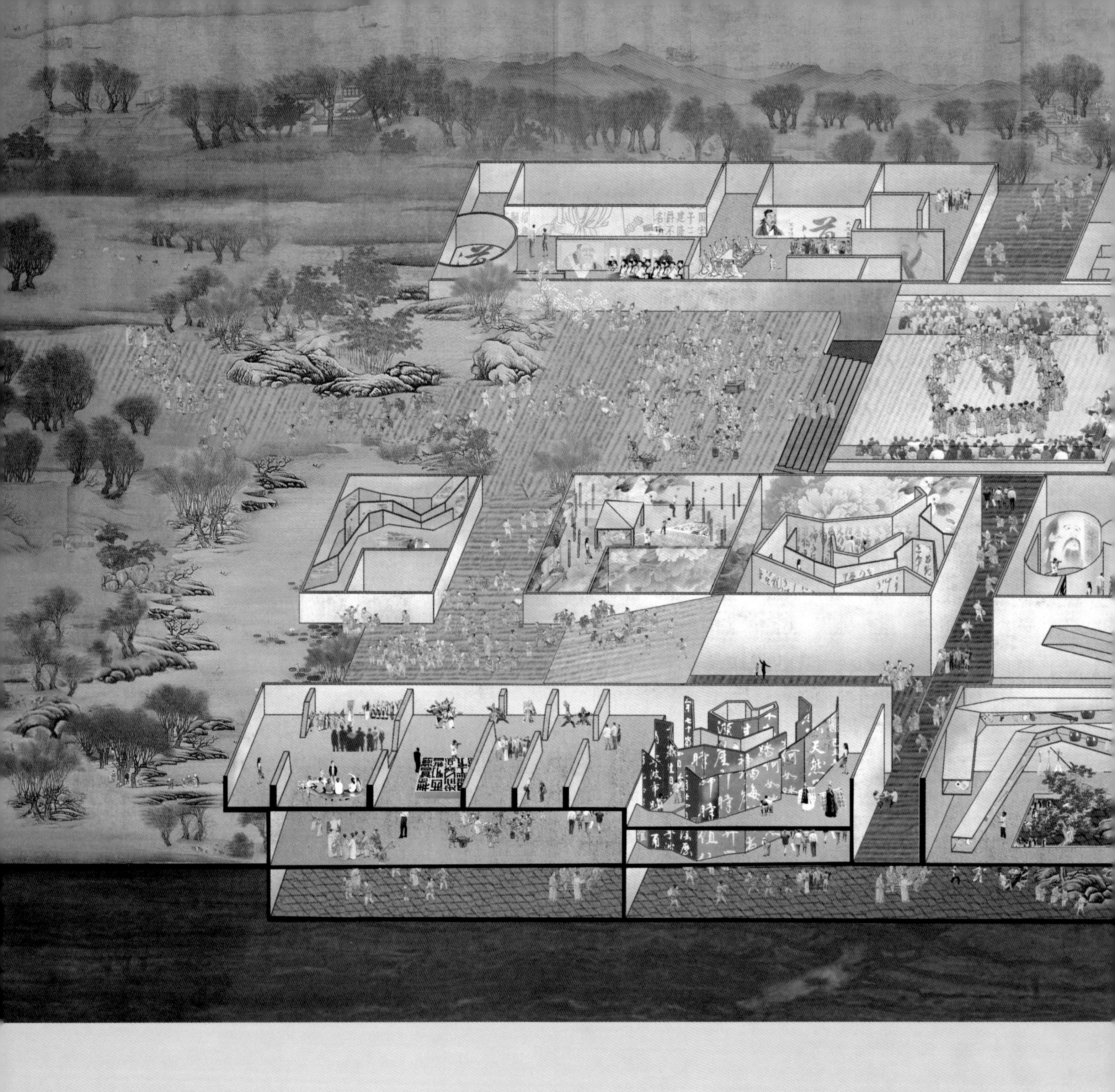

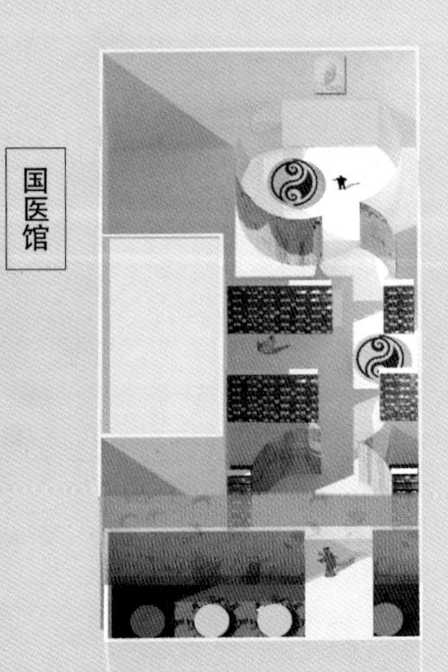
国医馆

戏曲馆

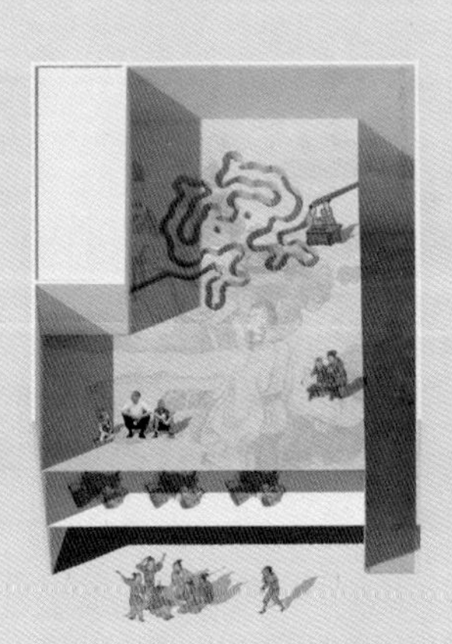
茶艺馆

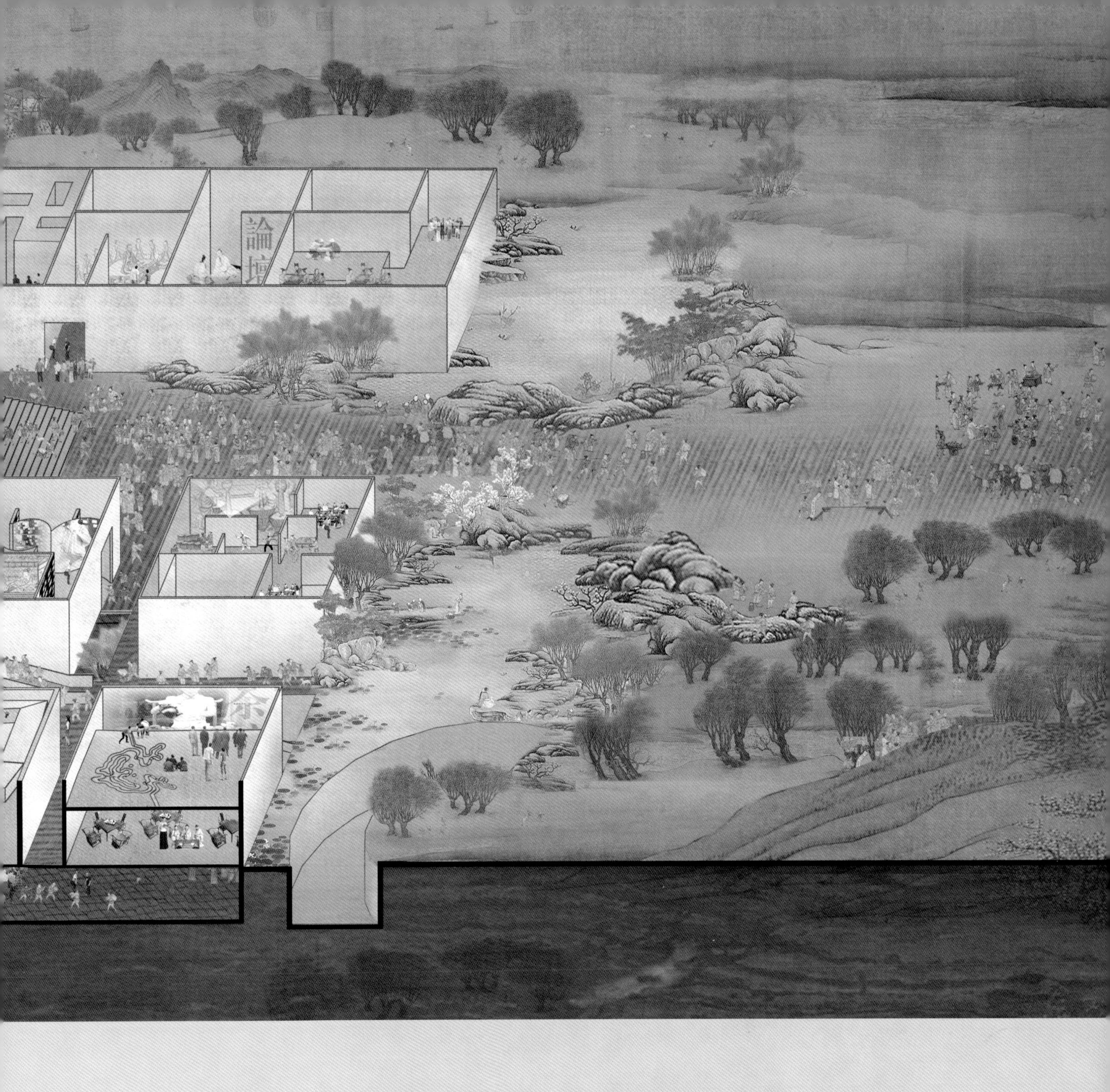

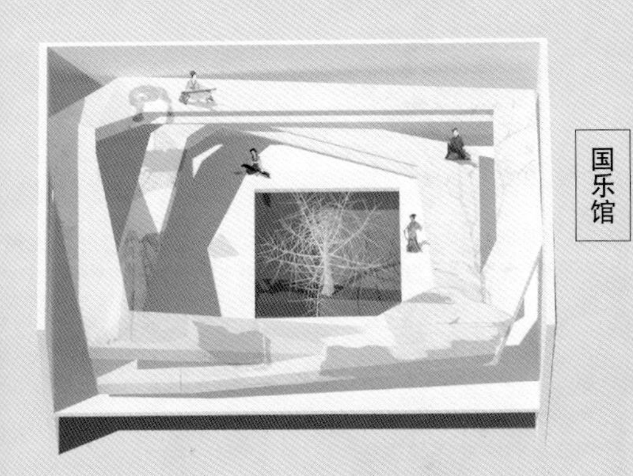

国乐馆

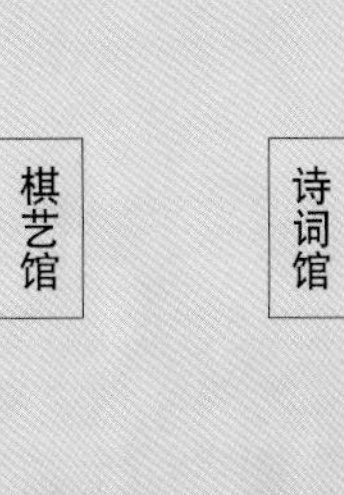

棋艺馆

诗词馆

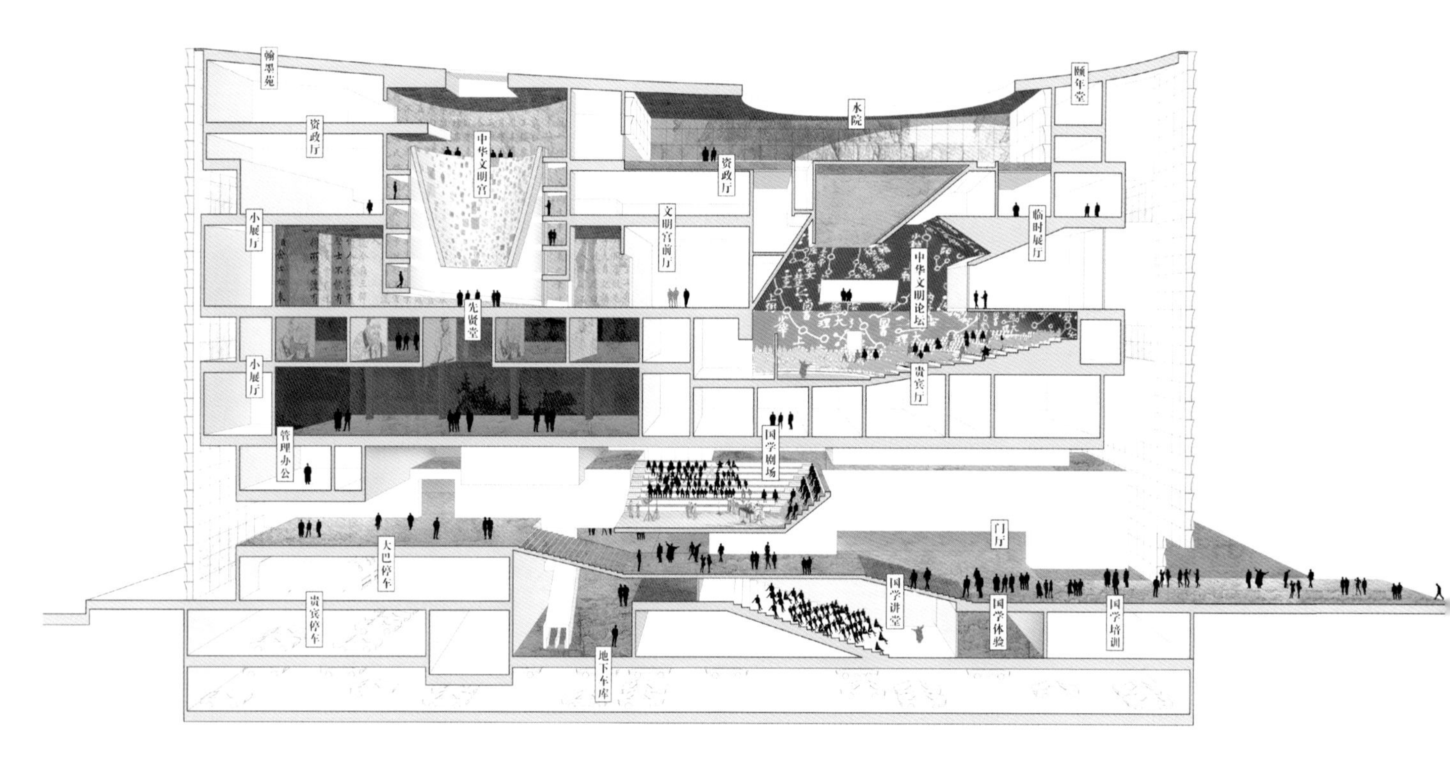
翰墨苑
水院
颐年堂
资政厅
中华文明宫
资政厅
小展厅
文明宫前厅
临时展厅
中华文明论坛
先贤堂
小展厅
贵宾厅
管理办公
国学剧场
门厅
大巴停车
贵宾停车
国学讲堂
国学体验
国学培训
地下车库
a-a 剖面图

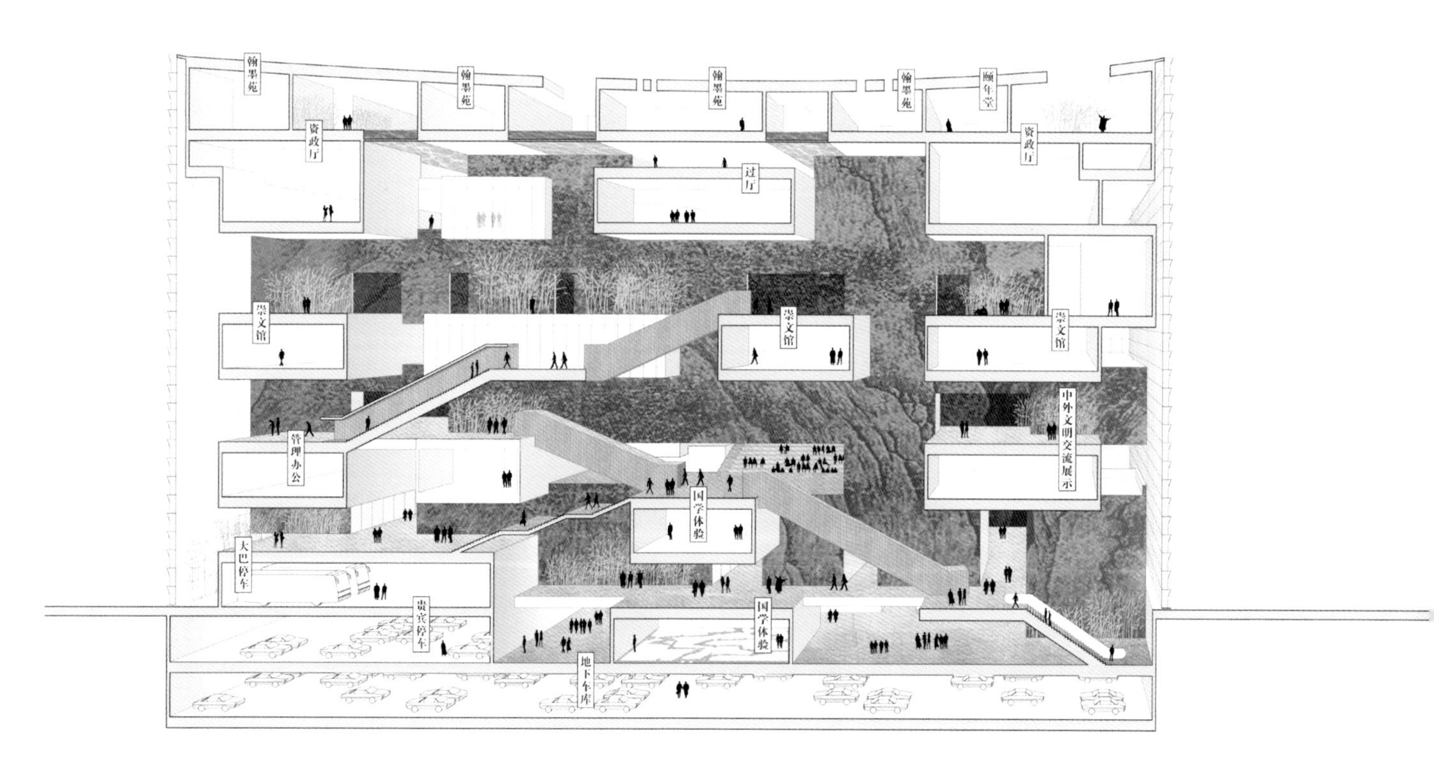
翰墨苑
翰墨苑
翰墨苑
翰墨苑
颐年堂
资政厅
资政厅
过厅
崇文馆
崇文馆
崇文馆
中外文明交流展示
管理办公
国学体验
大巴停车
贵宾停车
国学体验
地下车库
b-b 剖面图

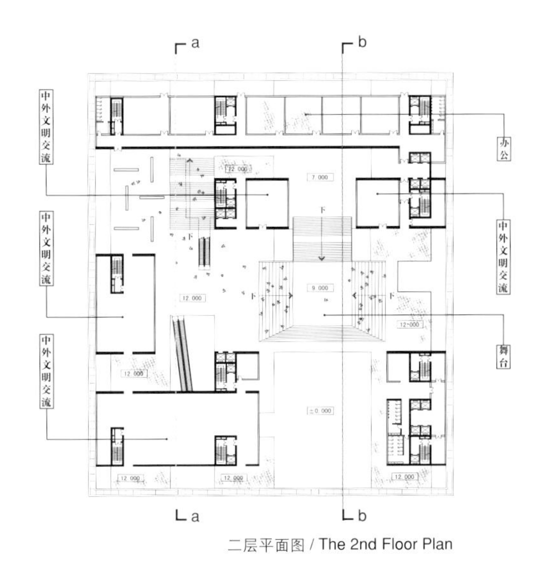

二层平面图 / The 2nd Floor Plan

四层平面图 / The 4th Floor Plan

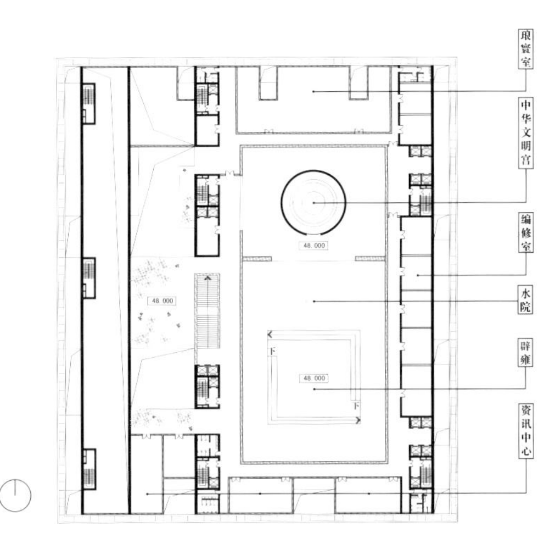

七层平面图 / The 7th Floor Plan

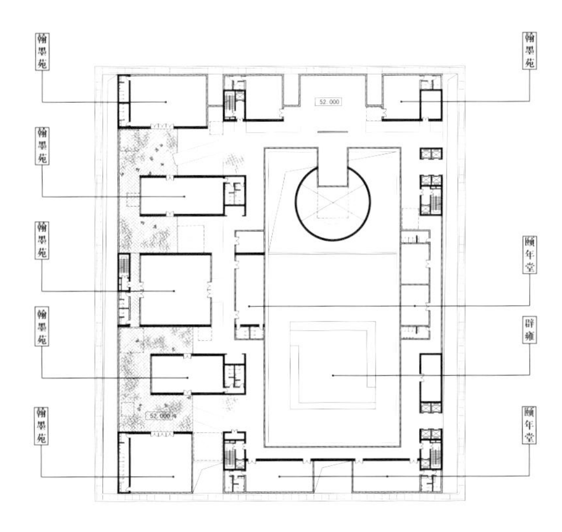

八层平面图 / The 8th Floor Plan

曹妃甸国际生态城文化中心　唐山

Caofeidian Culture Center, Tangshan Hebei Province

设计时间 / Design：2009 至今
用地面积 / Site Area：94600m²
建筑面积 / Building Area：70000m²
项目组 / Design Team：王辉 | 吴文一 唐康硕 | 赵洪言 孔祥磊
林秀清 曾皓 杨楠 李淳
合作 / Collaborators：中冶京诚工程技术有限公司（施工图）
业主 / Client：唐山市城乡规划局

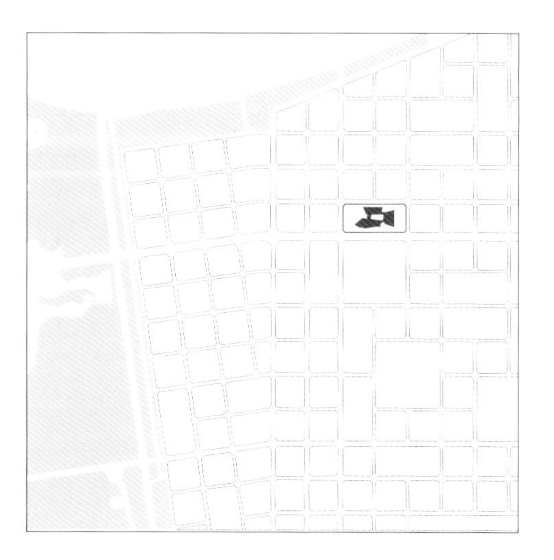

曹妃甸国际生态新城建在一望无际的滩涂上。在一个速生的城市中，任何建筑物的发展都失去了参照物。作为城市中最重要的公共建筑之一，文化中心显然是市民集体记忆的场所，不能因周边环境未知的变化而不断地“变脸”，也不能因为时尚的变迁而不断地“化妆”，因而设计采取了“内向布置的聚落”的策略，使其外围形式自然化，能应对周边的不可预知的种种变化。

文化中心的内容是由四大板块构成的，功能复杂的图书馆、活动中心和影剧院被设计成三个独立的山体，从而大大简化了它们之间错综复杂的矛盾。而将美术馆的内容放大为一个顶部的创意天街，同时可以俯瞰整个城市。

文化中心的形象是市民集体精神的表征。山的造型不仅仅是对山水城市的一种图示，也是对曹妃甸新城浪漫气质的一种诠释。底层三栋主体建筑围合而成的公共广场是城市事件的发生器，表演、集市、聚会、展示等等都可以在这里尽情地发生。 在不同的季节、不同的时段，这个广场可以为城市安排丰富多彩的活动。周边的三个建筑与这个空间不断地交流，并为这个广场提供有效的支持。顶层的天街使人能够在当代都市的喧嚣中找回小城镇的记忆，以及领略到时尚生活的魅力。有了这种空间关系，建筑物的造型反而低调化、中性化，更凸现其文化气质。

The Caofeidian International Ecological City Cultural Center is a public facility located in a vast expanse of wetlands soon to be a new city. In such a rapidly-growing city, the development of any building can hardly find a fixed reference because the building environment is unpredictable. As one of the most important public buildings in the city, the cultural center is clearly a place for collective memory. It cannot be constantly face-lifted according to changes in the surrounding environment, nor to the shifts in fashion trends. The design must be an introverted settlement with the image of natural forms of landscape, so as to be immune from any surrounding influence.

Three of the four parts of the cultural center are designed into three individual mountain forms, including a library, an activity center, and a theater. Such a design greatly simplifies the complexity of the sophisticated programs. The art gallery is highlighted and placed on top of the three mountains to form a sky-walk where one can overlook the entire city's growth.

As the icon of the city's collective spirit, the cultural center's mountain forms are not just an imitation of the landscape, but also an interpretation of the romantic aura of the time. Surrounded by the three buildings, a public square that acts as an event generator for performances, fairs, shows, etc. In different seasons and times of day, the flexibility of the plaza provides a variety of activities for the city. The buildings on the perimeter serve the square, as well as function effectively with their well-planned programs. The sky-walk above brings a cozy small town experience within the charms of a grand contemporary urban scenery. With this spatial configuration, the building is non-imposing, but holds a distinguished cultural character.

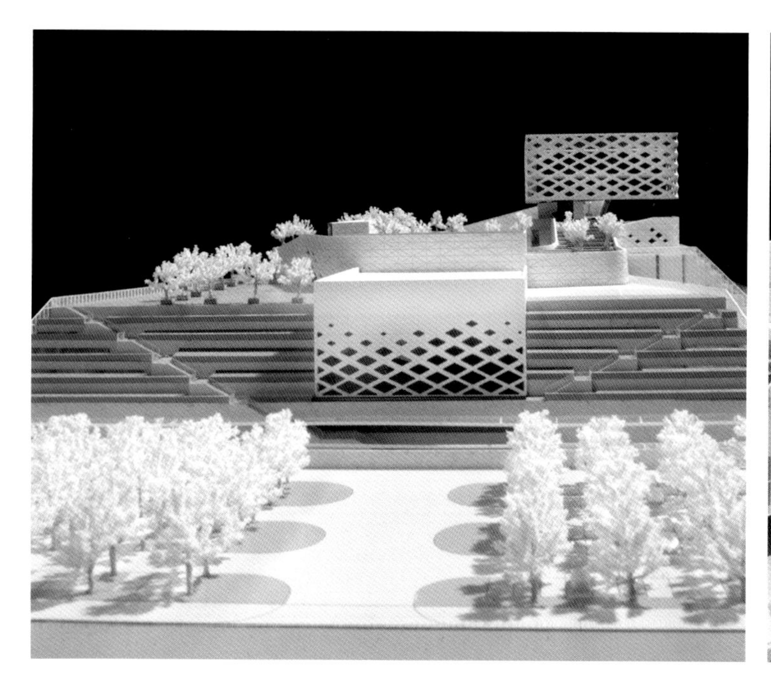

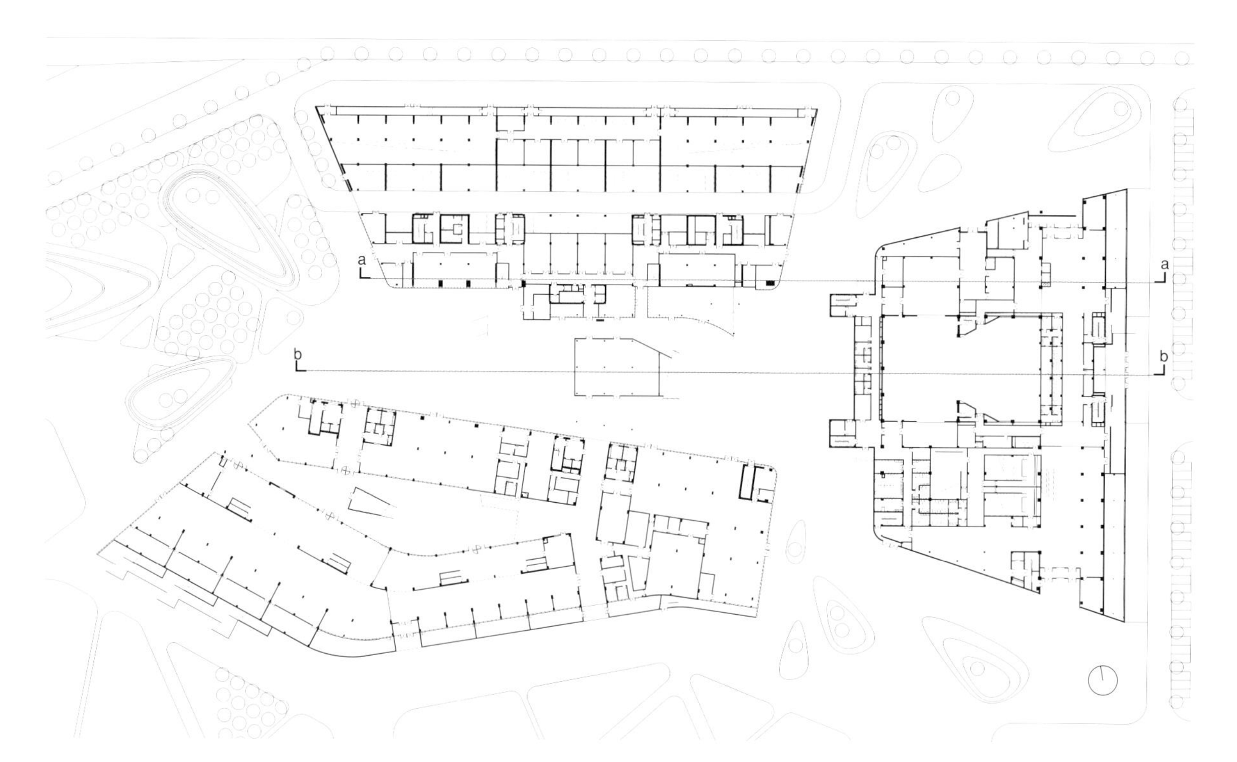

一层平面图 / The 1st Floor Plan

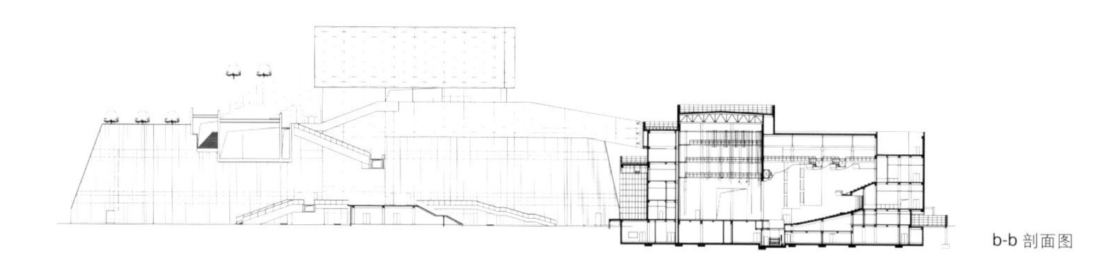

b-b 剖面图

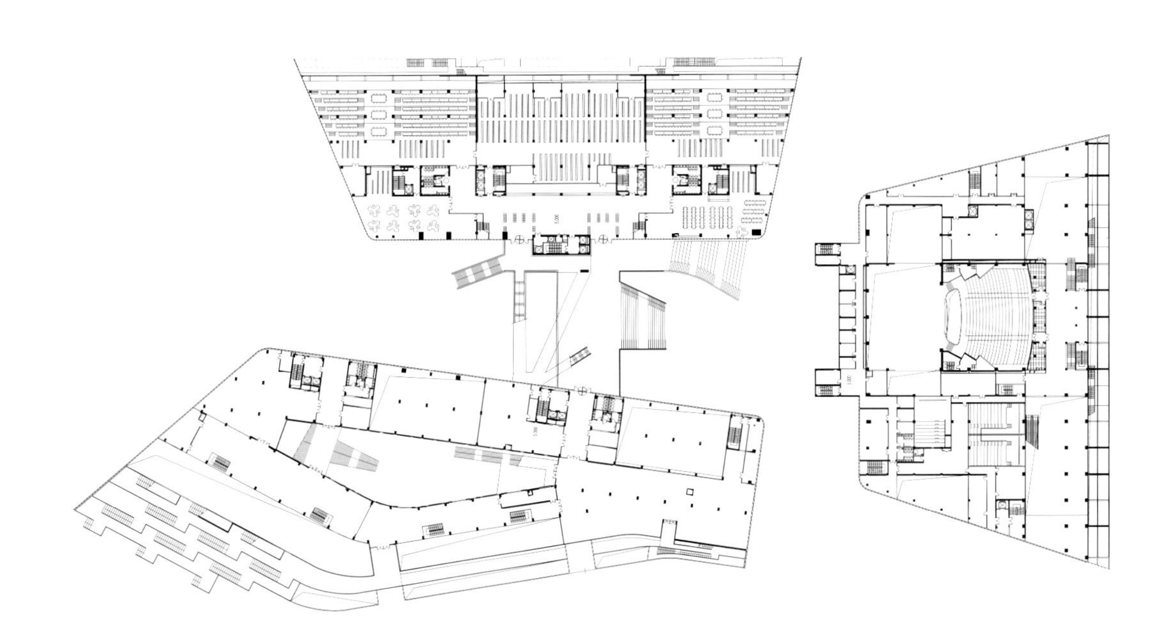

三层平面图 / The 3rd Floor Plan

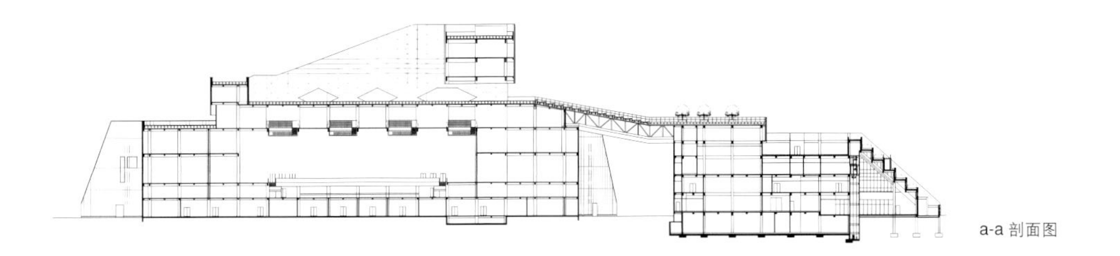

a-a 剖面图

香港中文大学综合教学楼

Integrated Teaching Building in Chinese University of Hong Kong

设计时间 / Design：2007
用地面积 / Site Area：13700m²
建筑面积 / Building Area：40200m²
项目组 / Design Team：孟岩 刘晓都 | Matthias Wolff 傅卓恒 | 杨期力 李晖 丁钰 Joshua Roberts
业主 / Client：香港中文大学

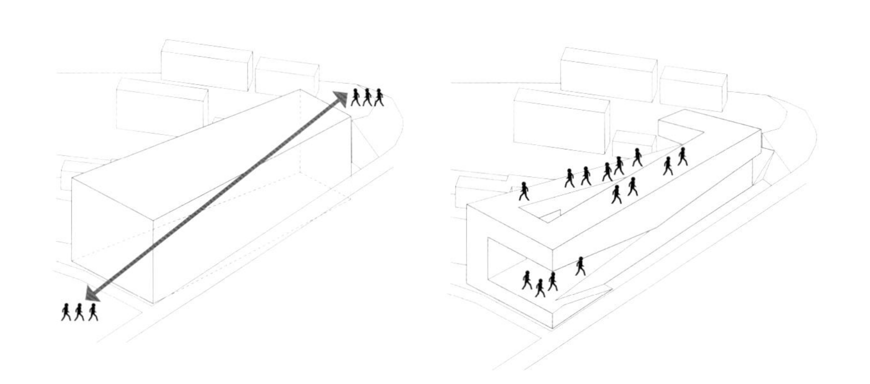

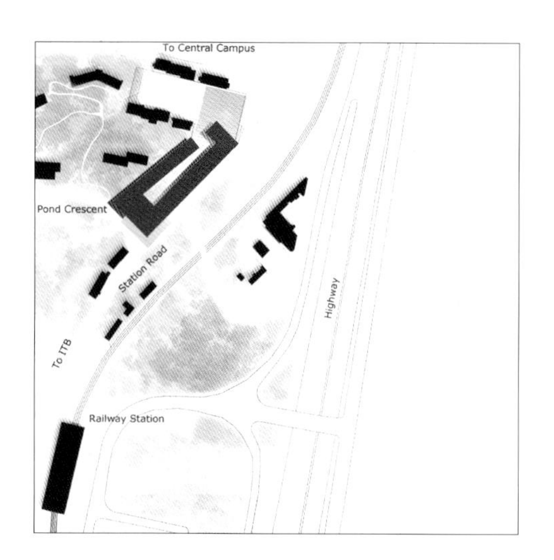

一个有气质的大学建筑应当有创造力、有个性、有开放的社区精神，它的设计应因地制宜。

该项目位于香港中文大学校门附近的一个落差很大的坡地上。山丘顶上有另一个书院。这个教学楼的建设，是一个能很好地连接上下校区资源的机会，给师生提供一个穿越、停留、交往、学习的复合空间。这一题解产生了一个三明治式的夹层布局：中间层作为开放的步道可以拾级而上，通过几个公共广场到达山顶；上层是图书馆、教室和院系教师办公室；下层则利用高差生成大型的阶梯讲堂。相对于自然的丘陵环境，超大的建筑体量也因此被弱化：下层部分成为基座，与山体相融合；上层架空部分水平展开，呈漂浮之态。平切的立面则意图突出整体感，使整个建筑达到坚实而空灵的气质，同时暴露出建筑的剖面。在建筑外面可以看到人们在台阶通道上下穿行，给这座教学楼带来动静交织的活跃氛围。一系列开放轻松的公共交往空间会成为大学生喜爱的活动场所，充满活力，并是一个理想化的校园。

本设计在竞赛评选中取得第一名，得到广泛好评。但之后因故没有成为实施方案。

Good university design should demonstrate the spirit of a creative and open community with special characteristics. In this competition project for Hong Kong Chinese University College Building design, designers respected the natural topography to formulate a site-specific spatial organization, endowing the building with an active character and focusing on the connectivity and the dynamic bonding between the space and the users.The site is located adjacent to the Hong Kong Chinese University gate, on a steep sloped hillside. The project strives to establish a dynamic linkage between two parts of the campus. Based on the building layout, the design concept emphasizes layers of mezzanines, creating a complex compound space, which encourages vertical pedestrian movement whilst providing visual connections between academic centers and informal learning spaces. The sandwiched layout introduces a main open central plaza, with stepped pathways along the topography ascending towards the peak via a series of fragmented small public plazas. Libraries, classrooms and college offices are located on the top level of the building, while the lower level of the building is used as the main stepped lecture hall conforming to the slope of the street.A solid pedestal foundation, which mingles with the mountain as a background, defines the volumetric character of the design. Floating above the lower level, horizontally-lifted interwoven mezzanine space fragments and moderates the mega-scale of the building. The sheer flat façade has enveloped the building to achieve unifying effect. The building is present as a showcase of educational activities, with direct visual connection of students' movements from the outside and flexible informal learning spaces from the inside. The project elaborates a series of weaved casual public spaces within which students can explore chances for social activities, academic lectures, casual communication and leisure.

The design was elected as the first place in the competition and received positive feedback. However, due to various reasons it was not constructed.

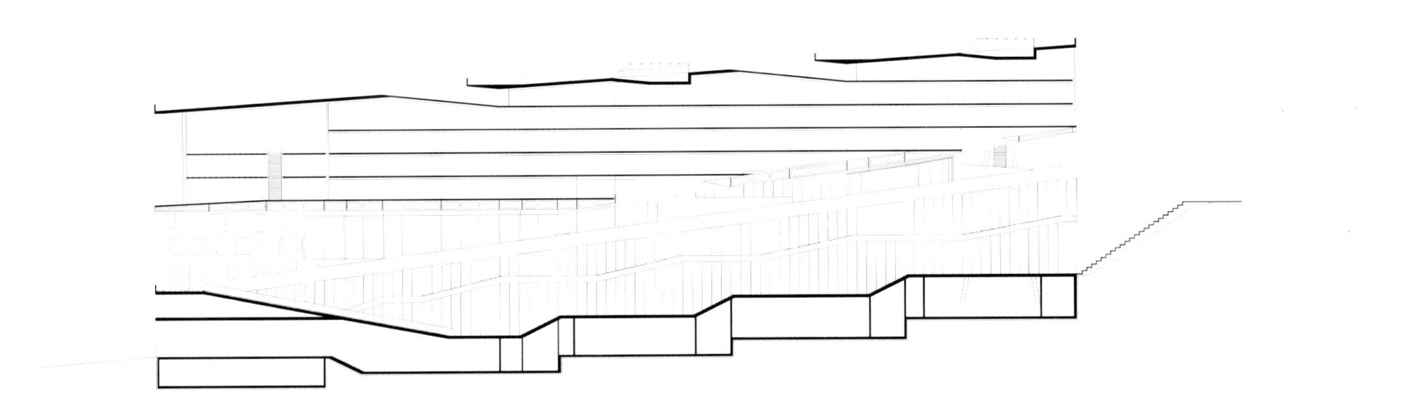

通道层平面图 / Level 3_Street Plan

LV 北京旗舰店立面设计

Façade Design of Louis Vuitton Beijing Flagship Store

设计时间 / Design：2011

项目组 / Design Team：王辉 | 吴文一 | 李淳 Jennifer H. Ha Deborah A. Richards 郑明璐 徐晶

合作 / Collaborators：Markus Schneider（Thismedia）

业主 / Client：Louis Vuitton Malletier

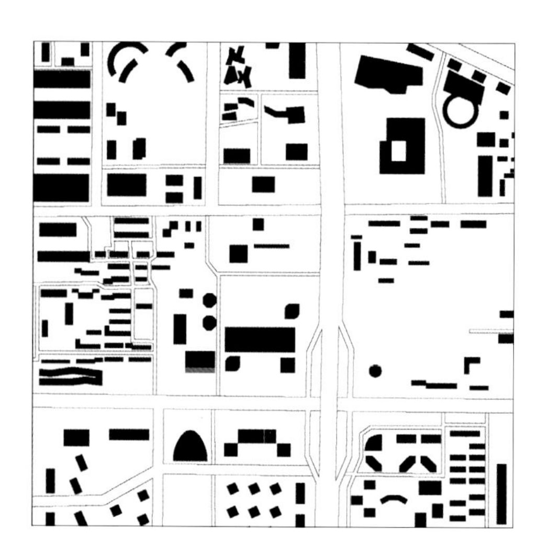

随着中国亦即将跃居为世界上第一位高端奢侈品消费国，时尚奢侈品牌LV也将其北京旗舰店定位为其在世界上最前卫的门店，并邀请五家中国建筑事务所参加设计提案。

这个店面选择在长安街与东三环交界的国贸中心，店面为两层高的建筑裙楼。设计抓住了店面沿长安街约有70米长展开面的特点，充分发挥这一水平长卷的优势，展示LV品牌高雅的气质和她迷幻的风格。

在千禧的第一个十年里，LV店面设计使用了平行的、有白色花格图案的双层玻璃，产生迷人的视幻觉。它成了这个品牌的符号。新旗舰店的设计目标是寻找新的展现这个符号的方式，使大众阅读既有认同感，又有新奇感。设计策略是：首先图案从二维转为三维，用一系列向街道发散、不平行的玻璃板，产生立体的图案叠加方式，并有移步异景、远近高低各不同的视觉效果。在这个基础上，设计立足点将其从一个私营店面转化为公共的城市装置，把格状的图案类比为城市的视网膜，通过与之相关的感应器来捕捉城市视觉信息，反映到藏在背景墙的灯光装置上，从而形成与城市互动的照明效果。

这个设计理念事实上是个可以普遍推广的原型，它可以成为千禧的第二个十年里LV的新形象。

China has grown to rank among the world's top consumers of high-end luxury goods. Louis Vuitton plans to build the world's most prestigious flagship store in its Beijing's location. The company invited five leading Chinese firms to participate in its design proposal.

Situated at the junction of Chang'an Avenue and East Third Ring Road, the store occupies the two story podium of the China World Trade Center. Designed to catch attention along Chang'an Avenue, the store has a seventy-meter long façade front. The design takes advantage of this long façade to illuminate the elegancy of the LV brand and its mirage-like style.

In the first decade of the new millennium, the LV storefront employed a design using a double-layered patterned glass to give a charming illusion. This two-dimensional illusion has become a recognizable symbol for the brand. This design strategy is to find a new way to re-present the brand, not only to make the brand immediately recognizable by the public, but also encourage a new dimension of curiousity. Our design method is to first translate the Louis Vuitton pattern from a two-dimensional to a three-dimensional reading. By placing an array of glass panels with LV's checkered pattern, and varying the angle of each panel, the façade will have a cinematic visual effect. Such a visual experience is intensified through the movement of different viewing points. This approach is not merely beneficial for the sake of the storefront but also makes for an interesting public installation. With a system of sensors installed behind the glass panels, now the checkered pattern on the façade also acts as the city's retina, as it captures the movement of the city and reflects this information into its lighting system to make an interactive lightshow for passersby.

The design is a prototype that can be adapted to future LV stores in the coming decade.

双层灯光装饰墙系统 /
Illuminted Cavity Lighting System

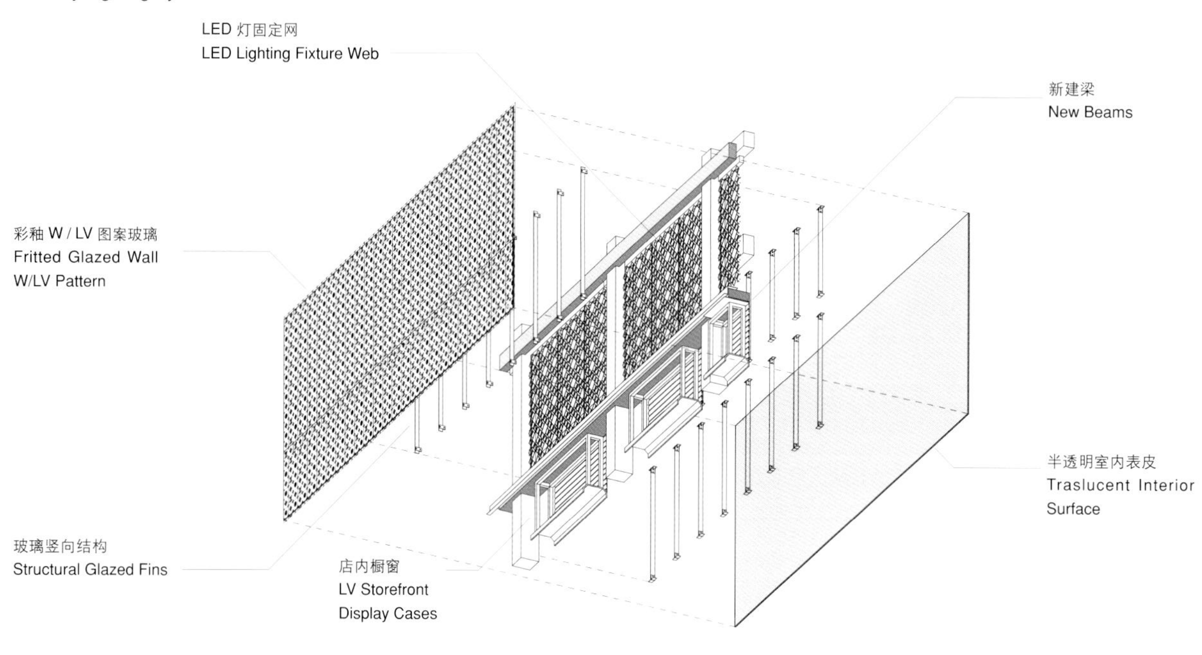

图案悬挂板系统 /
Graphic Hanging Pane and
Structural Hanging System

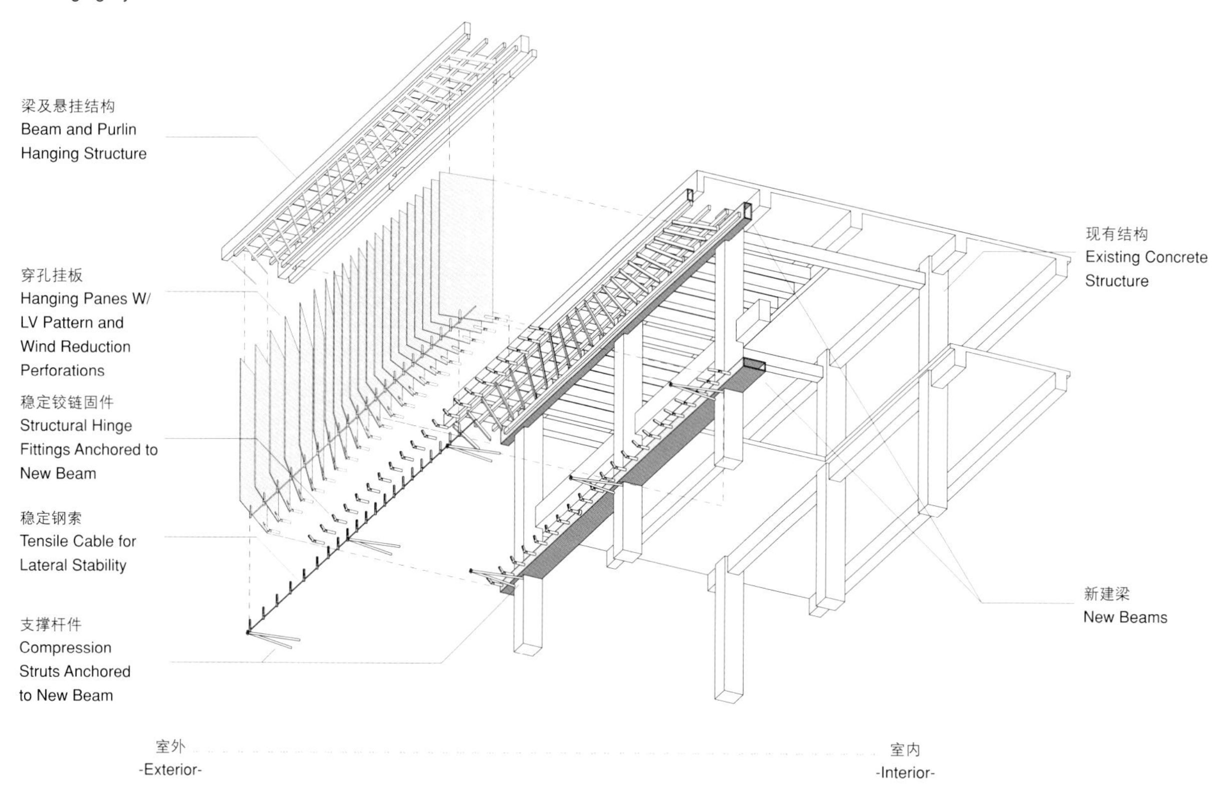

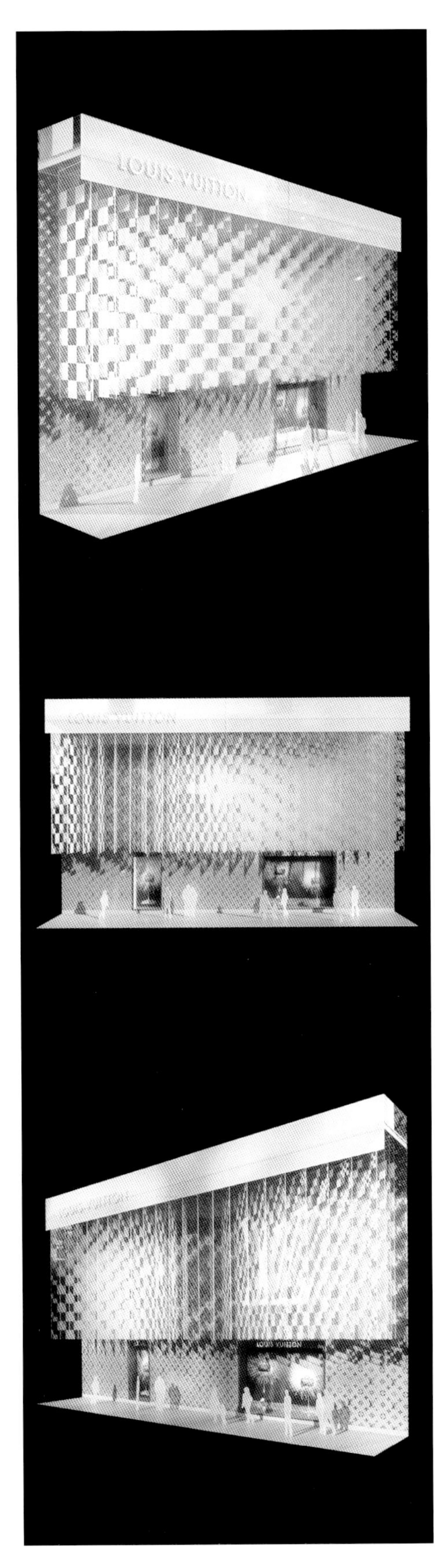

LOUIS VUITTON
LOUIS VUITTON

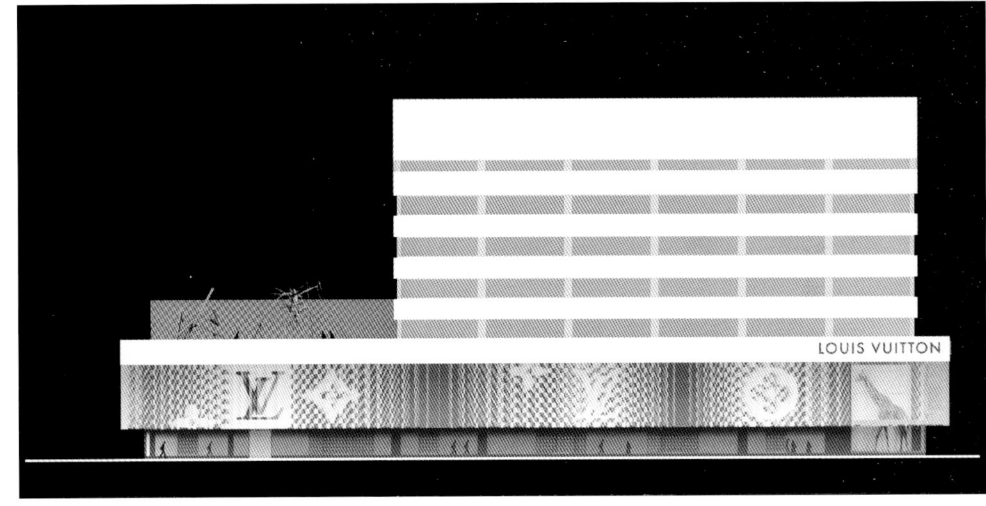
LOUIS VUITTON

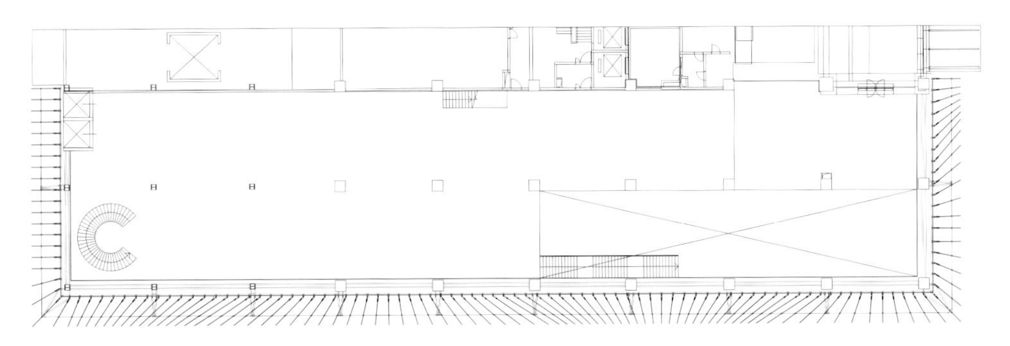

2010 上海世博会深圳案例馆

Shanghai Expo 2010 Shenzhen Case Pavilion

设计时间 / Design：2009
建成时间 / Completion：2010
建筑面积 / Building Area：600m²
项目组 / Design Team：孟岩 刘晓都 王辉 | 朱加林 | 林怡琳 饶恩辰 黄艺宏 陈兰生 汪源 | E-6 空间
策展合作 / Curatorial Collaborators：周红玫 牟森 高广健 杨勇 THISMEDIA
展品合作 / Exhibits Collaborators：龚志成 由宓 雷磊 张达利 余海波 何煌友
出版物合作 / Publication Collaborators：姜珺 史建 | 吐毛球
工程合作 / Construction Collaborators：华森建筑与工程设计 洪涛装饰 珠海晶艺玻璃 中建钢构
分展场活动合作 / Sub-venue Exhibition Collaborators：滕斐 华·美术馆
业主 / Client：深圳市人民政府 深圳市规划和国土资源委员会
摄影 / Photographer：杨超英 孟岩

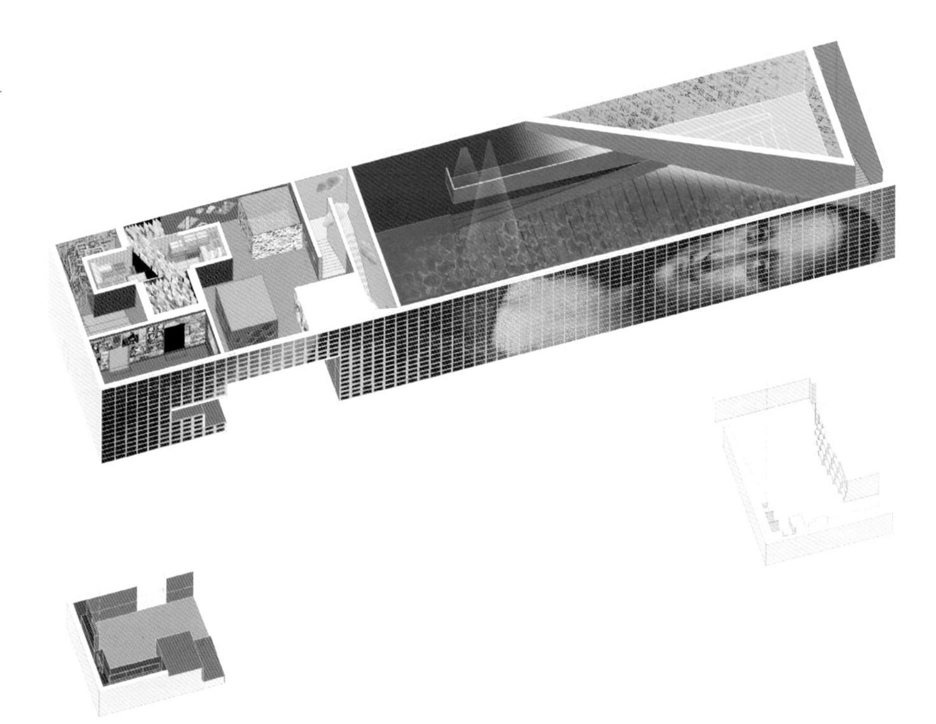

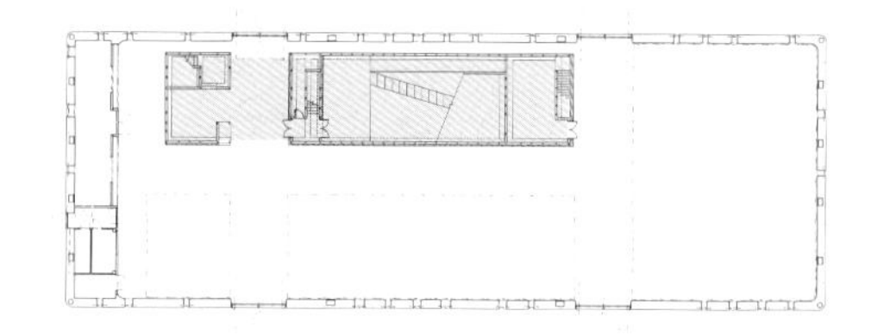

深圳在最后关头取得了世博会的入场券。都市实践作为总策展及展览设计机构，与市政府有关部门、合作艺术家一起在短短几个月时间内完成了这一高难度的任务。展览引入当代艺术、实验剧场、多媒体互动等多种艺术手段，与展览空间设计有机结合，颠覆传统的展示观念，使深圳案例馆既是展场，又是剧场，同时与深圳大芬村现场互动，实现从上海到深圳双城双展的空间跨界。

该展厅位于城市最佳实践区（UBPA）B3-2 展厅内部，面积为 390 平方米。展览设计为相对封闭的体验式展区，整体呈现为一个长 43 米、宽 9 米、高 7 米的大型剧场空间装置，从外到内形成丝丝入扣、高潮迭起的叙事结构和戏剧性体验。馆内空间与戏剧合谋构建经典叙事结构“三部曲”。借鉴德国音乐家瓦格纳提出的“总体艺术”概念作为空间叙事的出发点和基本理念，注重场所整体体验。分为序曲“大芬丽莎 / 深圳创世纪”，第一部曲“大芬制造”，第二部曲“大芬转型”，第三部曲“城市剧场：深圳，中国梦想实验场”及尾曲等部分。空间戏剧的叙事载体由大型油画装置、多媒体影像装置、城市采样体验区、村史文献装置以及空间剧场装置等一系列串接的、充满创意的体验区构成。整个展览空间本身作为一个主作品，其上有机附着众多相对独立又相互串联的作品。观者体验从 999 块手工油画拼合的“大芬丽莎”，过渡到活跃跳动的多媒体互动城市信息矩阵，再经过密集的强化个人体验的“村”中迷宫，最后抵达宏阔的城市剧场主厅，阅读到深圳这座年轻城市 30 年发展奇迹背后的原动力，感受深圳人的梦想和这座城市的精神。

Shenzhen was invited to the World Exposition 2010 in Shanghai at the last minute. URBANUS, as the curator and exhibition designer for Shenzhen's pavilion, finished this task in an extremely intensive schedule by collaborating with government authorities and artists. The Shenzhen Pavilion was designated as an installation showcase and theatrical stage. It aimed to exhibit contemporary art installations with an experimental theater, and to use interactive multimedia technology to seamlessly merge different exhibition spaces in an unconventional way.

The 390-square-meter pavilion was an enclosed cubical space with a rectangular footprint, 43 meters in length, and 9 meters in width with a 7-meter height ceiling. Visitors are guided to walk along the designated route, which is organized following the "trilogy" way. The design of the space was inspired by the German composer Richard Wagner's concept of Gesamtkunstwerk (total work of art). The "trilogy" of the architectural space is based on the rebuilding of the existing urban village in Shenzhen called "Dafen Village". The journey, in a chronological order, starts with a Prologue implying the Genesis of Shenzhen with a Mona Lisa façade – later known as "Dafen Lisa" – which was constructed as a montage of 999 hand-painted oil paintings units. The journey carries on to the second episode called "Dafen Transformation", with a multi-media installation and miniature landscape models showing urban sample pieces of the village. Then the journey leads to the third episodes called "Shenzhen, Frontier for China's dreams". It consists of numerous documentary films, historical prints and achieves, and ends with an epilogue with opportunities for review, discussion and contemplation. The pavilion itself has sublimed into a total work art linking among and emerging from thousands of independent artworks, enchanting a condensed labyrinth of an urban village, exploring the past 30 years of drastic development of Shenzhen, and at the same time, questioning the ambitions and the visions of this emerging miracle city.

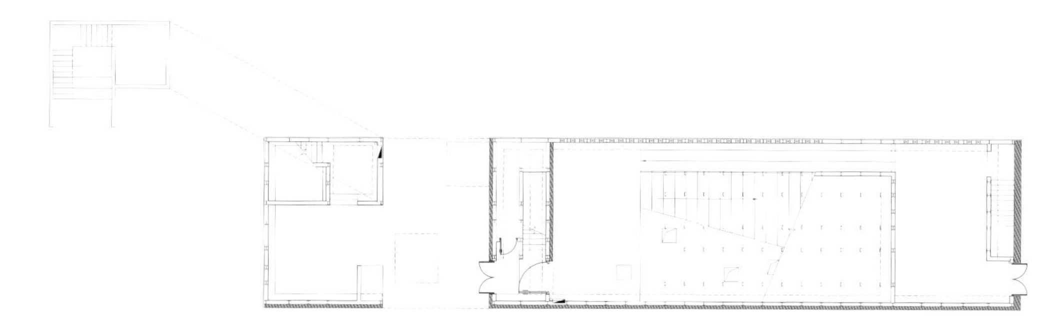

一层平面图 / The 1st Floor Plan

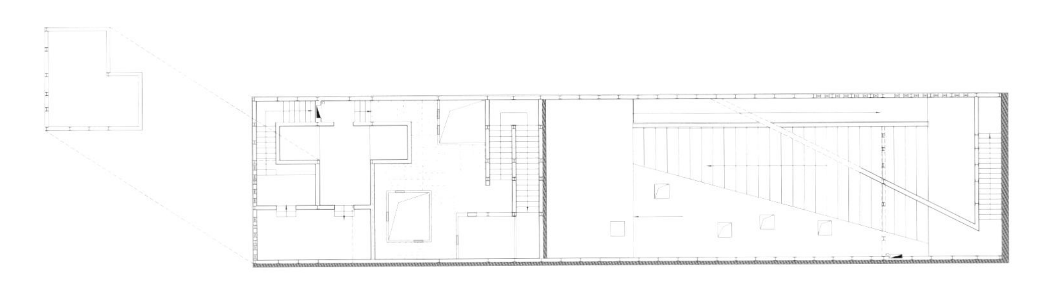

二层平面图 / The 2nd Floor Plan

都市实践访谈 刘晓都 + 孟岩 + 王辉

Interview

采访人 / 黄元炤

北京 2011.8.29

黄元炤：在都市实践的早期，主要着重于关注城市中常被忽略的待发展空间，并以介入城市空间的方式去解决现实问题，借此回应中国当代城市中的复杂性。以深圳地王城市公园第一期与第二期城市开放空间改造为例，用当代景观整合场地周边现存各自独立、支离破碎的闲置空地，形成一个新的公共空间，启发不同的使用方式与体验。这让我想到 OMA 大都会事务所在 1983 年巴黎拉维莱特公园竞赛中的尝试，当时库哈斯提出在没有任何单体建筑的参与下，创造出一系列的连环事件，并将已知的设计条件、空间与预知可能发生的内容同时组织在场地里，随意的配置，体现一种开放性与灵活性相结合的设计。这种连续性的叠合系统突破了既有传统的街道与广场形式。你们如何看待在设计操作上与库哈斯的差异性？建筑本身需对变化有着包容性吗？

孟岩：URBANUS 都市实践一个主要特点，或者说考虑问题的方式，是从问问题开始。我们做一件事，要先知道这个地方到底有什么问题，然后针对问题找适宜的解决办法。建筑师想干什么事就要干什么事，是很危险的。深圳在近十来年内，以极昂奋的发展速度，积累了许多大型项目。当这个新兴城市的发展在数量与质量上呈现不均衡状态时，我们看到它极度欠缺一种对于城市空间质量、尺度、内容和文化上的关怀与再思考。我们的起步是始于城市开放空间改造。起因是，当时正值世界公园年，深圳被评选为国际花园城市之一，这背后的重要意义相当于今日的大运会或世博会，说明国际社会在关注亚洲的新兴城市。对都市实践来说，这是一个契机，去观察和设计城市如何发展自己的自明性及生活方式。都市实践为此提出了公共空间的设计是城市设计，而不只是景观设计，不只是限于某一个公园，它的影响应涵括至城市范围。有计划性地把城市中的公园相互串通，形成一个由微观叙事构成的宏大故事。这好比一部由 36 集构成的电视连续剧，是由一系列小故事、分故事组合成的一个好看的大故事，而不是一个乏味的故事。从地王城市公园系列后，我们长期持续地关注并设计了不下十几个公共空间。对都市实践而言，其理念是在这个拥有高密度硬体的城市中，植入它所缺乏的微小、软化、贴近生活的空间。

回到你提及是否受到其他理论的影响的问题，理论研究确实开阔了我们的思考深度，但我们有自己的见解。对于每个具体项目，都市实践都会提出有针对性的策略和手段。"都市造园"计划不只是在做一个个公园，而是叙述一个个不同的策略：在密集的城市中散置许多微小的空间，使人在城市漫游时不经意地邂逅不同主题的故事；在一个大公园里面切分了很多场景，当你路过的时候，一系列的场景会随之变化，就像看一部电影一样，进而让人与空间发生互动。

王辉：我们中国设计师当今的实践是否总在某个西方理论的阴影下？这个问题暗含了一个古旧的说法，即解决中国的城市问题，需要植入西方的理论与经验。事实上，中国乃至亚洲当前新的社会、经济、人口条件，必然会产生新的城市问题，会面临着全新的挑战，从而创造了新的知识与经验。虽然西方的知识可以借鉴，但应变之道始终应立足于自身的问题。

黄元炤：从城市观点介入建筑方面来谈，城中村的改造是都市实践一直所关注的课题。我观察到都市实践对应不同项目有着不同的策略与手法，有提出共存、拆除、填充、缝合以及加建或局部改建的各种可能性，有以商业介入并试图形成高密度混合的多功能商业村落，更有提出城中村中创造出多元城市活动的村中城的口号，是否将商业置入也作为设计时考虑的重要因素？

刘晓都：回归到对城市的认知，混合是个必然的观念。库哈斯对城市的见解也基于强调混合带来的效应。孟岩与王辉有在纽约生活多年的经验，深刻体会到大城市的形成和运作机制，再以纽约的城市认知直接地比照到深圳这种正在迅速转变中的城市生活状态，有助我们去预测这城市未来的走向。城中村改造的普遍手段是拆除现有形态，将其置换成所谓的花园小区。这种模式的后果是显而易见的。我记得当时最早有争议性的城中村，正好就位于罗湖区地王公园边，所有的主流舆论都提倡将城中村拆除，却忽略了城中村有良好的城市社区中所缺乏的活力这一价值。都市实践尝试以一种微观的城市视角，去关注和保护城市的异质边缘与另类活力。

孟岩：城中村现象是中国当代城市中普遍的难题，现今中国的城市有求大、求美、求同、求纯的趋势，其价值观与城中村存小、存丑、存异、存杂相背道而驰。脏乱、密集、有自然活力的区域成为不被城市允许和认可的附属物，成为被改造对象。这种现象，不仅存在于北京和深圳，更在中国的任何城市。

刘晓都：这是种理想化的城市想象，而不可能成为城市的基本现象。

孟岩：现在追求过度的美学和视觉化，也许能替城市带来干净的环境或是符号式的地标，如鸟巢、中轴线、长安街等等，这是城市发展中不可避免的趋势。但是我们通过城中村认知到，一个好的城市要有相当的包容性，需要去容纳很多异质的东西，它必须同时供给有钱的人和贫困的人、好人和坏人以同等的居住及使用公共空间的权利。城市作为一个居住的容器，需回应社会各个阶层的需求，不能屈从于形式上的趋同化。深入到外表看来脏乱的城中村里面，会发现它提供的 24 小时的便利服务、丰富的各地小吃、多元的人际关系网络，是很难被一种规矩的形态所取代的。这就给我们提了个醒，城中村虽然确实存在着脏、乱、差的负面表象，但有着无限的活力。反之，高级的门禁社区里提供了赏心悦目的景观、干净奢华的设施，然而却使城市割裂，城市人的生活变得畸形。这再度验证了城市的构成是需要互相补充与交织的，而非完全的排他性。

黄元炤：我观察到你们对城中村有一种尊重低层与现实存在的态度，试图以贴近当地居民的视点，并以一个直观、直白的概念去面对城市。这让我想到了现象学，它同样是以直观的方式，去掌握事物本质的过程，排除掉所有预想与成见，本质是观察对象所呈现的内容进而推想出来的各种可能。对于你们在设计前期时，是否也直观的切入社会学或者是人类学的角度，来判断场地应拆迁、保留或置入的各种可能性呢？我想听听你们对这方面的看法？

孟岩：在当今的世界里面，建筑属于少数的几样东西能够明确地跟一个地点发生关系。过去，事件的发生都跟地域存在着一种密不可分的关系，现在人跟地域则处于逐步丧失联结的状态，好比今天我来深圳，明天去北京，后天在美国，不被土地束缚。这现象不仅反映在人，也反映于衣食住行上，例如人人都用的 iPhone，搞不好会经常拿走边上人的手机。对我而言，建筑跟 iPhone 有着程度上的差别，建筑应当跟特定地点有某种特殊的联系。我们始终坚信建筑不应该失去地域性，我从来不认可一个建筑师的作品可以被任意地以同一形式被建造于任何地点。在这个世界上，建筑作为极少数能长期占据城市空间、并留下持续发展轨迹的文明形式，应当表明自身的态度及立场。至于建筑本身搬迁去留问题，则为更深层的考虑了。让建筑跟土地、跟周边的环境及人文能够直接或间接地产生关系与衔接，保有外形与内在的联系，这是我们所追求的愿景。不是说我们每一个作品都达到了这个目标，但这是我们的愿望。

王辉：设计的亮点在于能否在建筑跟设计条件（如场地）产生的必然性关联中，敏锐地洞察出设计问题、场所精神及使用者需要，并有效地反映在应对策略中，从而提出一个有意思的解决方案来。当前建筑师极需要这种能力，能直观地去淬炼出问题的本质，让设计成果成为对设计问题的个案化的解答，使设计方案有自明性。面对中国的超速发展，要一下子解决几代人积累的问题，必须要用快速和成功的模式来运作，所以我们看到了市场上各式各样成熟的标准户型，这都是当下立竿见影地解决快速需求的产物。这种标准答案式的、有效的模式，使所有的城市与区域驱同化。作为一个设计师，想既要获得大众的共同认可，又能保留自我，需要一种努力。在这个层面上，当今世界上那几个活跃的大师的共通点，就是做每一个项目的时候，即便和其他的项目在空间处理手法上呈现出的表面相似性，但思考的关注点是不一样的，这才能彰显他们的自明性和创造性。

黄元炤：您说的自明性和创造性似乎也体现在大连海中国美术馆项目里。就我的观察，这个项目带有所谓地景式建筑的含意，以仿似地表起伏状态的连续性空间来探讨建筑与自然地景相互的关系，让强调流动性与连贯性自然形态融入到建筑空间中，我想听听您如何直观的去看待这个倾向于概念性的设计？如何去解决这个地貌与内部相对应的空间？

王辉：我们总是试图针对问题提出相对比较适宜的解答，这需要做出正确的判断。海中国美术馆，实际上是为一个上百万平方米的社区所设的销售中心。这个项目先行于整体开发，因此我们着手设计时，基地处于未开发的状态下，场地周边没有一个参照物，如果将它设计成一个普通的房子，并不合理。因此做了个地景式的建筑。另一个设计的出发点，是试图创造出人和建筑之间的互动，人能来来回回地在空间里上上下下迂回和停留。在这个过程当中，让人和房子自然地产生一种互动的关系，使消费者对建筑的喜爱自然地延伸到购房的欲望。从这个角度看，人与建筑和环境之间的亲密关系的建立，也巧妙地转化为一种营销方式。

黄元炤：你们在近几年有一个明显的转向或者说是尝试，关注到更多当代流行的建筑语言，比如表象性。值得一提的是，北京都市实践与深圳都市实践的作品似乎得暂时拆开来看，两方在设计的实践上不论是表现手法还是关注点都有着一定的差异性。比如北京的都市实践，在辽宁海中国美术馆中关注到如何在一种人与房子，人与环境之间产生互动关系，更倾向于地景式建筑的体现，在白云观珍宝花园则回归到传统与现代的命题，用现代玻璃材料去体现一种表皮性的设计语言，企图创造古代文化的再现。换言之，北京的都市实践更贴近于当代或是国际主流，而深圳的都市实践从早期关注城市、环境、人文开始，以景观手法切入城市公共空间、城中村的改造到近期的土楼公社，更贴近于都市脉动或者是本土。你们是如何看待自己的差异性？是地域上的不同所导致吗？还是个人有不同的追求？

王辉：在这十年里，确确实实面临了许多的有关价值的判断，包括了刚才所提的城中村，当时主流社会强调的是提倡清除它，而都市实践却站在主张保护的立场上。这理念对于我们深圳或北京办公室来说是没有区别的。在北京这个有深厚历史文化底蕴的都市，应加倍思考如何看待城市历史及文物。虽然北京现在设有历史保护片区，但也带来了规划上的故步自封，导致表面化的保护，不能妥善地利用城市资源，扼杀了历史遗存的现实意义，使城市最终驱向了无聊、中庸和落后的困境。以天津鼓楼街改造项目和北京白云观珍宝花园设计为例，天津老城厢由于修复了一个假鼓楼，带来了整个街区的复古。这种应景却不应时的做法，带来沉重的历史包袱，使鼓楼街商业很萧条，无法供给城市新的经济刺激与活力。因而，在我们的规划中，对这种复古的做法勇敢地说不。同样地，白云观珍宝花园的设计，是用当代的语言与古典语境产生的对比的美感去响应社会脉动，透过改造白云观的边界环境，把原先被城市遗忘的角落重新找回来，并发挥其存在的价值。这项目特意用工业化条件下生产的现代玻璃幕墙体系与经过千百年沉淀下来的手工建构体系作对比，让两个同样精美的体系能够协调，用当代的材料与技术勇敢地去与历史对话。这是做这层表皮的立意，而不是去表达时尚的信息。

黄元炤：就我的理解，URBANUS 都市实践的公司精神里面提到努力继承现代主义先驱理想，那么我想听听你们如何解读对于现代主义的追求？

孟岩：我强调我们追求更多的是现代主义的理念，尤其是现代主义中对人文的思考，我们并没有想简单继承现代主义的形式和风格。

黄元炤：你们是倾向站在现代主义人文理念的那一个角度来切入设计？

孟岩：现代主义的理念和实践在美国与欧洲之间存在着差异性。欧洲早期的现代主义，源起于工业建筑，以很多现代主义先驱为工人做的住宅为例，以技术性及功能性的方式去响应普通老百姓的需求，为平民弱势着想，并主张社会平等的理念，形式上着重于功能合理性，并无任何多余的装饰，从这基础上才产生出现代主义的思潮。不幸的是，在 20 世纪 30 年代，现代主义被正式介绍到美国以后，它转变成时尚和资产阶级的象征，现代主义是权贵人士才有资格享用的。它丧失了起初欧洲现代主义早期先驱们的理想社会意识与左派的平等意识，渐渐剥离掉了精神层面与文化上的内涵，而只留下了形式的空壳。比如要想盖一个美术馆，玻璃盒子已失去其所代表的深层原意，仅是一种非常光鲜的表现。其实对于现代主义的形式，我们不注重它是否为玻璃盒子，它是否功能第一，我们并不完全遵循现代主义建筑的某些原则。我们所提倡的是现代主义中的社会理想性、革命性的实践，那种试图利用建筑改变社会的意念。也许最后现代主义不免俗地会沦落成资产阶级的玩物，但我们坚信在中国快速发展的机会下，建筑师仍然可以有现代主义者那种改善社会的影响力。

黄元炤：回应你们说的对于现代主义的理解，如何回归到现代主义中的对人文理念的坚持？你们是站在对于人性关怀的角度来思考设计，还是着重于周边环境的呼应？

孟岩：关注当代的问题，当代的人。

黄元炤：拿深圳大芬村美术馆作一个城市建筑的案例，首先关注到周边环境的人及油画生产方式，在设计中设立机制让美术馆跟通俗艺术进行对话与结合，积极地与城市中的当地居民或者是外来的游客产生对话，意图与城市纹理、市井生活取得联系，借以解决当下社会所存在的问题，这个部分你们有何看法？

王辉：现代主义这个名词包含了不同的概念，不同人对现代主义的论述也不同。我们所表述的现代主义更是一种精神意志，而不是外表的形式，这使它与其他表征风格的主义有本质的区别。有趣的是，这一百多年的历史中，从来没有任何一个主题与思潮像现代主义这样有如此鲜活的生命力，它的革命精神是不灭的，但形式是灵活的。这也包括当代艺术与现代艺术之间本质性的区别。两者表面形式上虽接近，但是深层有不同的关注点。而我发现它们关注的方式是一样的，就是关注当下。现代主义实现了多数人要求社会进步的需求，我们认同这种精神与理念，却不具体去复述现代主义的手法、方式，或者说是风格，我们更认可延续现代主义的精神层面，用批判性眼光，勇于介入社会，去解决问题。建筑师的任务不仅仅是盖房子，而是建设一个社会。

孟岩：十多年前 URBANUS 都市实践成立时写下的中心思想就是"努力继承现代主义先驱理想"。经过这十几年的变迁，都市实践坚信的精神与理念从来没有变过。现代建筑师究竟在面对城市化、社会的尖锐矛盾等等问题时该如何定位自己？我们认为建筑师不应只限于自己专业的建筑学的建构领域里，在这个大时代里，要走出建筑的小圈子，将投入社会与城市作为当前最重要的一件事。不应只一味地沉浸于新的形式语言和空间趣味，尤其当面临这个后工业和信息时代，必须去理解社会中隐含的多元因素，并且做出准确的判断，才不致随波逐流。

刘晓都：从面对这些大问题或者从作品当中的角度去找寻设计出路，我觉得还是很重要的。针对刚才谈到现代主义的理念，你所谓的人文交流，我认为更倾向于都市实践强烈的平民意识。认同平民意识，深入平民思想，关注平民生活。我们选择不将自己剥离出社会的大多数，坚定地跟社会同步。

黄元炤：URBANUS 都市实践在下一步发展中，有什么样的计划?

刘晓都：首先是从过去十年的实践中总结经验教训，进一步思考中国当代社会和设计行业与世界的关系和特质，进而想办法找出下一个十年的方向。当然，我们仍然是实干派，愿意去尝试和承担风险。我们在保持事务所已有的设计实力和南北互动格局的同时，力图拓展新的方向和尝试新的模式。第一个努力是打造符合和支持建筑创作的技术平台。都市实践联合几家相似的事务所共同成立北京互联盟建筑设计公司，建设自己的施工图团队。第二个尝试是成立研究部，名为 URBANUS Research Bureau（URB），进行城市研究和城市设计。第三个举措是成立香港分部，以海外青年建筑师为主体的设计研究小组。第四个计划是在深圳成立梧桐山城市研究工作坊，建立珠三角城市研究的国际合作平台。URBANUS 都市实践力图在未来的中国城市发展的浪潮中发挥更大的作用。

作品年表

Chronology of Works

文化与教育建筑 / Cultural & Educational Building

大连海·中国美术馆
Maritime Art Museum

地点 / Location：大连 / Dalian
设计 / Design：2007
建成 / Completion：2008
用地面积 / Site Area：7654m²
建筑面积 / Floor Area：4216m²
项目组 / Team：王辉 | 陶磊 | 赵洪言 杜爱宏 郝钢 张永建 刘爽 陈春 张永清
合作 / Collaborators：大连都市发展建筑设计（施工图）
业主 / Client：华润（大连）有限公司

深圳前海展示厅
Shenzhen Qianhai Exhibition Center

地点 / Location：深圳 / Shenzhen
设计 / Design：2010 ~ 2011
建成 / Completion：在建 / Under Construction
用地面积 / Site Area：20979.6m²
建筑面积 / Floor Area：4347.0m²
项目组 / Team：孟岩 | 姚殿斌 | 周成松 林挺 | 陈兰生 张天欣 谢盛奋 李嘉嘉 陈丹平
合作 / Collaborators：中国建筑科学研究院深圳分院（施工图）
业主 / Client：深圳市前海深港现代服务业合作区管理局
深圳市万科房地产有限公司（代建总承包）

北京国家远程教育大厦投标
National Center for CET & CCR & TVU

地点 / Location：北京 / Beijing
设计 / Design：2010
用地面积 / Site Area：11700m²
建筑面积 / Floor Area：110000m²
项目组 / Team：王辉 吴文一 | 郝钢 | 林秀清 Jennifer H. Ha
张淼 Deborah A.Richards 杨涛 杨楠 彭喆
业主 / Client：教育部远程教育大厦建设办公室

唐山隆达陶瓷文化馆
Tangshan Bone Porcelain Pavilion

地点 / Location：唐山 / Tangshan
设计 / Design：2010
用地面积 / Site Area：10405m²
建筑面积 / Floor Area：16000m²
项目组 / Team：王辉 吴文一 | 魏燕 | 林秀清 郑娜 高大智 陈宣儒
业主 / Client：唐山隆达骨质瓷有限公司

深圳观澜版画基地美术馆及交易中心
Art Museum and Trading Center for Guanlan Print Base

地点 / Location：深圳 / Shenzhen
设计 / Design：2009
用地面积 / Site Area：31000m²
建筑面积 / Floor Area：29405m²
项目组 / Team：孟岩 刘晓都 | 何勇 胡志高 | 潘志国 熊嘉伟 张丽娜 黄艺宏
业主 / Client：深圳市宝安区人民政府

南方科技大学图书馆
Library of South University of Science and Technology

地点 / Location：深圳 / Shenzhen
设计 / Design：2010 ~ 2011
建成 / Completion：在建 / Under Construction
用地面积 / Site Area：8627m²
建筑面积 / Floor Area：10727m²
项目组 / Team：孟岩 | 姚殿斌 | 苏爱迪 林怡琳 | 陈春 | 黄志毅 王俊 朱伶俐 谢盛奋 李嘉嘉 吴锦彬 陈兰生 王彦峰 罗俊辉 吴少文 | 刘爽 李图
合作 / Collaborators：深圳市建筑科学研究院（施工图）
业主 / Client：深圳市建筑工务署 南方科技大学建设办公室

龙岗区三馆一城方案设计国际竞赛
Longgang Three Centers + Book City International Competition

地点 / Location：深圳 / Shenzhen
设计 / Design：2011
用地面积 / Site Area：24593m²
建筑面积 / Floor Area：63500m²
项目组 / Team：刘晓都 孟岩 | 林海滨 | 饶恩辰 温谨馨 陈兰生 罗仁钦 张新峰 孔倩珺 李耀宗 刘思钊 沈振中 孙艳花 臧敏 黄艺宏
业主 / Client：深圳市万科九州房地产开发有限公司

中国冶铁（鞍山）博物馆
Anshan Museum of Iron & Steel

地点 / Location：鞍山 / Anshan
设计 / Design：2010
用地面积 / Site Area：20000m²
建筑面积 / Floor Area：9000m²
项目组 / Team：王辉 | 吴文一 | 杨楠 曾皓
业主 / Client：鞍山市规划局

上海嘉定新城示范中学
Jiading New Town High School

地点 / Location：上海 / Shanghai
设计 / Design：2009 ~ 2011
建成 / Completion：2012
用地面积 / Site Area：25000m²
建筑面积 / Floor Area：67500m²
项目组 / Team：王辉 | 吴文一 刘银燕 | 魏燕 张淼 | 杜爱宏 张永健
陈春 刘爽 王菁菁 霍振舟 崔海峰 李佳颖
合作 / Collaborators：中外建建筑设计有限公司（施工图） 无界景观（景观设计）
业主 / Client：上海嘉定新城发展有限公司

甘肃新校园计划苇子沟小学
Weizigou Primary School New Campus Plan

地点 / Location：甘肃成县 / Cheng Xian, Gansu
设计 / Design：2008
建成 / Completion：2009
用地面积 / Site Area：4112m²
建筑面积 / Floor Area：1400m²
项目组 / Team：刘晓都 | 胡志高 张新峰
合作 / Collaborators：卓越置业集团有限公司（赞助）协鹏建筑与工程设计（施工图）
业主 / Client：甘肃省成县教育局

天津大学新校区会议中心及校前区设计
Tianjin University New Campus Conference Center and Main Gate Area

地点 / Location：天津 / Tianjin
设计 / Design：2011
用地面积 / Site Area：26600m²
建筑面积 / Floor Area：9000m²
项目组 / Team：王辉 吴文一 | 郝钢 | 赵洪言 孔祥磊 郑明璐 刘妮妮
业主 / Client：天津大学

南方科技大学体育馆
Gymnasium of South University of Science and Technology

地点 / Location：深圳 / Shenzhen
设计 / Design：2008 ~ 2011
建成 / Completion：在建 / Under Construction
用地面积 / Site Area：7075.32m²
建筑面积 / Floor Area：9733.76m²
项目组 / Team：孟岩 | 姚殿武 | 林怡琳 | 周娅琳 熊嘉伟 胡志高 陈兰生 艾芸 王彦峰 吴凡
合作 / Collaborators：深圳市建筑科学研究院有限公司（施工图）
业主 / Client：深圳市建筑工务署 南方科技大学建设办公室

成都华侨城欢乐谷剧场
OCT Mosaic Theater

地点 / Location：成都 / Chengdu
设计 / Design：2007 ~ 2008
建成 / Completion：2009
用地面积 / Site Area：6500m²
建筑面积 / Floor Area：3400m²
项目组 / Team：王辉 | 刘旭 | 孔祥磊 张永建 陈岚 郑晨亮
合作 / Collaborators：中国建筑西南设计研究院（施工图）
业主 / Client：成都天府华侨城实业发展有限公司

深圳湾公园小品及服务设施
Shenzhen Bay Park Pavilions and Facilities

地点 / Location：深圳 / Shenzhen
设计 / Design：2006
建成 / Completion：2011
用地面积 / Site Area：5000m²
建筑面积 / Floor Area：3700m²
项目组 / Team：刘晓都 | 徐罗以 | 邢果 张海君 | 李强 黎靖 夏淼
魏志姣 饶恩辰 左雷 黄艺宏 姜汶 林刁超 张烁 任晖
合作 / Collaborators：美国 SWA（规划）中规院深圳分院（规划）
北林苑景观设计（施工图）
业主 / Client：深圳市建筑工务署 深圳市土地投资开发中心

上海嘉定飞联纺织厂改造
Redevelopment for Feilian Textile Factory, Jiading

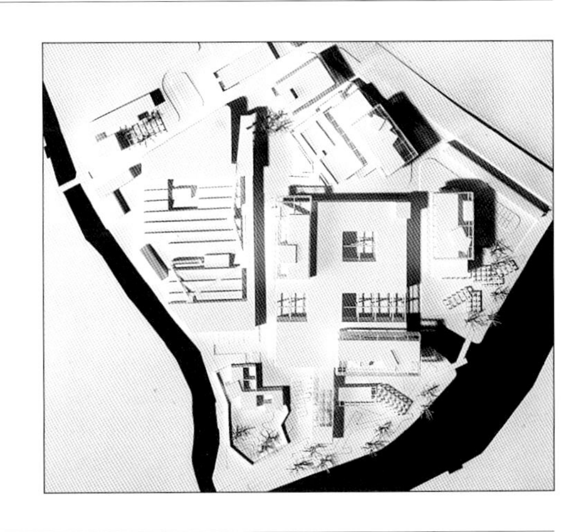

地点 / Location：上海 / Shanghai
设计 / Design：2011
建成 / Completion：在建 / Under Construction
用地面积 / Site Area：8409m²
建筑面积 / Floor Area：43137m²
项目组 / Team：王辉 吴文一 | 魏燕 唐康硕 王韬 | 赵洪言
成直 孔祥磊 郑明璐 徐晶 张哲
合作 / Collaborators：联创国际设计集团公司（施工图）
业主 / Client：上海嘉定国投公司

深圳大亚湾核服办公楼
Office Building of CGN Service Company in Daya Bay

地点 / Location：深圳 / Shenzhen
设计 / Design：2009
建成 / Completion：2011
用地面积 / Site Area：8255m²
建筑面积 / Floor Area：14000m²
项目组 / Team：孟岩 | 朱加林 姚殿斌 | 胡志高 林挺 | 魏志姣 | 左雷 何勇 黄艺宏 胡伊硕
合作 / Collaborators：中汇建筑设计　广州容柏生工程设计深圳事务所（结构）
天宇机电设计（设备）深圳市建筑科学研究院（施工图）
业主 / Client：广东大亚湾核电服务（集团）有限公司

深圳地铁车辆段综合楼与物资总库
Shenzhen Metro Division Building and Warehouse

地点 / Location：深圳 / Shenzhen
设计 / Design：2008 至今
建成 / Completion：在建 / Under Construction
用地面积 / Site Area：19160m²
建筑面积 / Floor Area：79612m²
项目组 / Team：孟岩 刘晓都 | 林海滨 | 魏志姣 | 张新峰 席江 罗仁钦 张震 艾芸 刘明 黄煦 李昊
孙艳花 吴春英 刘思钊 臧敏 潘志国 姚晓微 | 廖志雄 林挺 胡伊硕 于晓兰 刘洁
合作 / Collaborators：极尚建筑装饰设计（室内）北京中外建建筑设计深圳分公司（施工图）
业主 / Client：深圳市地铁集团有限公司

嘉定工人文化活动中心
Workers' Union Building, Jiading New District

地点 / Location：上海 / Shanghai
设计 / Design：2009 至今
建成 / Completion：在建 / Under Construction
用地面积 / Site Area：8410m²
建筑面积 / Floor Area：23863m²
项目组 / Team：刘晓都 | 姚晓微 李耀宗 | 魏志姣 | 苏爱迪 刘思钊 Manuel Sanchez-Vera
谢盛奋 胡志高 张新峰 刘明 姜轻舟 张英 | 于晓兰 刘洁 林挺
合作 / Collaborators：互联盟建筑设计（施工图） 佳艺图室内设计
业主 / Client：上海嘉定新城发展有限公司 上海嘉定区总工会

北京亦庄郎波尔工业园
Lampower Industrial Park

地点 / Location：北京 / Beijing
设计 / Design：2010
建成 / Completion：2012
用地面积 / Site Area：66000m²
建筑面积 / Floor Area：120000m²
项目组 / Team：王辉 | 吴文一 | 魏燕 张永建 段云龙 | 郑娜 林秀清 李湃
合作 / Collaborators：电子第十一设计研究院北京分院（施工图）
业主 / Client：北京亦庄郎波尔光电股份有限公司

上海多媒体谷
Shanghai Multimedia Valley Towers

地点 / Location：上海 / Shanghai
设计 / Design：2005 ~ 2006
建成 / Completion：2008
用地面积 / Site Area：7017m²
建筑面积 / Floor Area：21226m²
项目组 / Team：王辉 | 段云龙 | 魏燕 陈春 高林 张永清 骆王华 张淼
合作 / Collaborators：上海同建建筑设计（施工图）
业主 / Client：上海多媒体谷投资有限公司

深圳第一创业大厦
First Capital Building

地点 / Location：深圳 / Shenzhen
设计 / Design：2008
用地面积 / Site Area：4111m²
建筑面积 / Floor Area：35000m²
项目组 / Team：刘晓都 孟岩 | 何勇 熊嘉伟 李立德 黄艺宏
业主 / Client：第一创业证券有限责任公司

大中华 CEPA 广场
Greater China CEPA Plaza

地点 / Location：深圳 / Shenzhen
设计 / Design：2010
用地面积 / Site Area：12200m²
建筑面积 / Floor Area：230324m²
项目组 / Team：刘晓都 | 邓军 李耀宗 陈兰生 林挺 姜轻舟 饶恩辰
Michael H. Rogers Mares E. C. Marc
业主 / Client：大中华国际集团（中国）有限公司

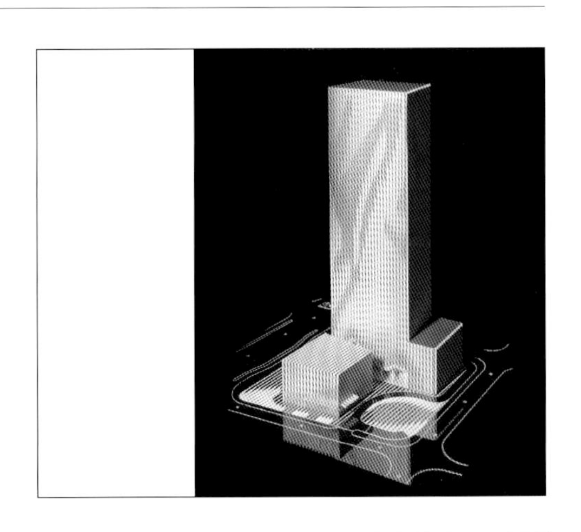

天津响锣湾 B-12 云滨大厦
CLOUD TOP Building

地点 / Location：天津 / Tianjin
设计 / Design：2008
建成 / Completion：在建 / Under Construction
用地面积 / Site Area：10015m²
建筑面积 / Floor Area：97719m²
项目组 / Team：王辉 | 郝钢 | 林秀清 李淳 胡美玲 戴天行 罗晶 裴雷
合作 / Collaborators：天津市建筑设计院九所（施工图）
业主 / Client：天津云滨置业投资有限公司

南山公安局出入境办公楼
Nanshan Exit and Entry Administration Building

地点 / Location：深圳 / Shenzhen
设计 / Design：2009
建成 / Completion：2010
建筑面积 / Floor Area：2687m^2
项目组 / Team：刘晓都 | 张新峰 | 梁广发 张丽娜
合作 / Collaborators：郭群设计（室内） 同济人建筑设计（施工图）
业主 / Client：深圳市南山区建筑工务局

鄂尔多斯 20+10 项日
Ordos 20+10

地点 / Location：鄂尔多斯 / Ordos
设计 / Design：2010
建成 / Completion：在建 / Under Construction
用地面积 / Site Area：120000m^2
建筑面积 / Floor Area：20000m^2
项目组 / Team：王辉 吴文一 | 吕琢 郝钢 魏燕 | 郑娜 杜爱宏 张永建 段云龙 杨楠 曾皓
合作 / Collaborators：互联盟建筑设计（北京） 联创国际设计集团公司（施工图）
业主 / Client：鄂尔多斯市政府

唐山湾伊泰广场
Tangshan Bay Yitai Plaza

地点 / Location：唐山 / Tangshan
设计 / Design：2011 ~ 2012
用地面积 / Site Area：36756m^2
建筑面积 / Floor Area：202150m^2
项目组 / Team：王辉 吴文一 | 郝钢 | 孔祥磊 郑明璐 徐晶 陈睿
合作 / Collaborators：北京中联环建文建筑设计（施工图）
业主 / Client：内蒙古伊泰置业有限责任公司

北京华夏基金数据清算中心
Data Auditing Center of China AMC

地点 / Location：北京 / Beijing
设计 / Design：2009
建成 / Completion：在建 / Under Construction
用地面积 / Site Area：9000m^2
建筑面积 / Floor Area：20000m^2
项目组 / Team：王辉 | 吴文一 | 唐康硕 | 魏燕 张永健 陈春 刘爽 郑明璐 陈睿
合作 / Collaborators：中冶京诚工程技术有限公司（施工图）
业主 / Client：华夏基金管理有限公司

深圳迅宝一期研发楼及专家楼
SUNBOW R&D Building and Expert's Lodge

地点 / Location：深圳 / Shenzhen
设计 / Design：2009 至今
建成 / Completion：在建 / Under Construction
用地面积 / Site Area：52746m²
建筑面积 / Floor Area：研发楼 15000m² 专家楼 5000m²
项目组 / Team：刘晓都 | 徐罗以 | 林海滨 张新峰 苏爱迪 | 王俊 刘思钊 陈兰生 黄中汉 孙艳花 黄煦
合作 / Collaborators：中咨建筑设计（施工图）
业主 / Client：深圳市迅宝投资发展有限公司

北京亦庄迅宝工业园
Beijing SUNBOW Industrial Park

地点 / Location：北京 / Beijing
设计 / Design：2011 至今
用地面积 / Site Area：26000m²
建筑面积 / Floor Area：58000m²
项目组 / Team：刘晓都 | 许小东 吴文一 | 朱伶俐 陈丹平 吴锦彬 刘洁 刘羽
合作 / Collaborators：互联盟建筑设计（施工图）
业主 / Client：北京迅宝新材料技术有限公司

坪山新区规划分局办公楼改造
PLC Pingshan New District Office Building Renovation

地点 / Location：深圳 / Shenzhen
设计 / Design：2009 ~ 2010
建成 / Completion：2010
用地面积 / Site Area：21399m²
建筑面积 / Floor Area：14956m²
项目组 / Team：刘晓都 | 朱加林 | 刘明 | 欧阳祎 左雷 孙艳花
合作 / Collaborators：郭群设计（室内）
业主 / Client：深圳市规划和国土资源委员会坪山分局

深圳东都大厦
Dongdu Building

地点 / Location：深圳 / Shenzhen
设计 / Design：2008 ~ 2009
建成 / Completion：2011
用地面积 / Site Area：10750m²
建筑面积 / Floor Area：20006m²
项目组 / Team：孟岩 | 朱加林 | 张震 | 李昊 艾芸 黄煦 孙艳花 吴春英 王俊 林海滨
合作 / Collaborators：深圳机械院建筑设计有限公司（施工图）
业主 / Client：深圳市龙岗区东都实业有限公司

华南城城市设计
Huanancheng Urban Design

地点 / Location：深圳 / Shenzhen
设计 / Design：2012
用地面积 / Site Area：270000m²
建筑面积 / Floor Area：2000000m²
项目组 / Team：刘晓都 孟岩 | 林达 | Travis Bunt　Kurt Franz 张天欣 胡润泽
合作 / Collaborators：深圳市新城市规划建筑设计（规划）
业主 / Client：华南国际工业原料城（深圳）有限公司

光明新区中央公园概念规划方案
The Conceptual Planning Of Centre Park, Guangming New Town

地点 / Location：深圳 / Shenzhen
设计 / Design：2008
用地面积 / Site Area：2.37 km²
项目组 / Team：孟岩 吴文一 | 邢果 | 王衍 廖志雄 张烁 任辉 黄艺宏
合作 / Collaborators：联盟设计（陈伟航、Peter L. Knutson、李川）
业主 / Client：光明新区管委会 深圳市规划局

昌平未来科技城核心区城市设计
Beijing Future S & T Park Central Area Urban Design

地点 / Location：北京 / Beijing
设计 / Design：2010
用地面积 / Site Area：1355000m²
建筑面积 / Floor Area：2200000m²
项目组 / Team：王辉 吴文一 | 成直 | 王韬　Jennife H. Ha
业主 / Client：北京未来科技城开发建设有限公司

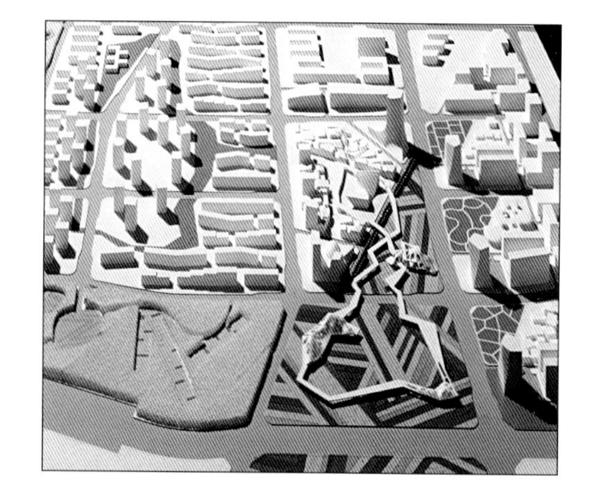

龙岗区大运新城南北区城市设计
Urban Design for Longgang Universiade New Town Southern and Northern Zones

地点 / Location：深圳 / Shenzhen
设计 / Design：2010 ~ 2012
用地面积 / Site Area：175000m² + 82000m²
建筑面积 / Floor Area：2982000m² + 1698000m²
项目组 / Design Team：刘晓都 | 林达 | Henry Martes　Michael H. Rogers　Alex Kong
Jenny Jo　Karin Lam　Paula Asturis 陈婧波
合作 / Collaborators：深圳市新城市规划建筑设计有限公司（规划）
业主 / Client：深圳市规划和国土资源委员会龙岗管理局

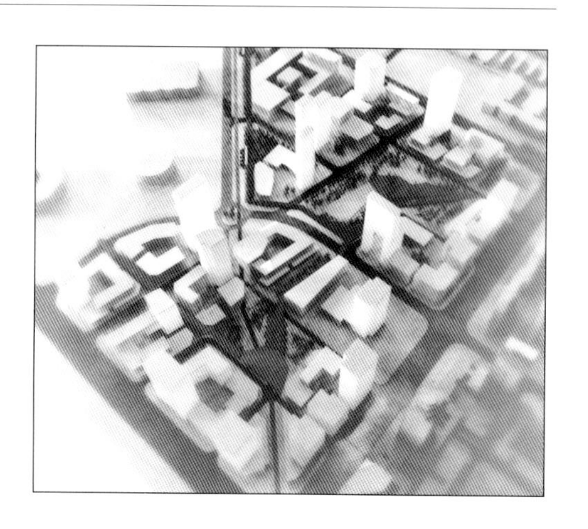

西涌鹤斗村保护规划研究
Xichong Hedou Village Preservation Planning and Research

地点 / Location：深圳 / Shenzhen
设计 / Design：2008
用地面积 / Site Area：110000m²
建筑面积 / Floor Area：64979m²（改造后）
项目组 / Team：刘晓都 | 王衍 | 欧阳祎 黄艺宏 孔繁锦
合作 / Collaborators：新城市规划建筑设计（规划）
业主 / Client：深圳市规划局滨海分局

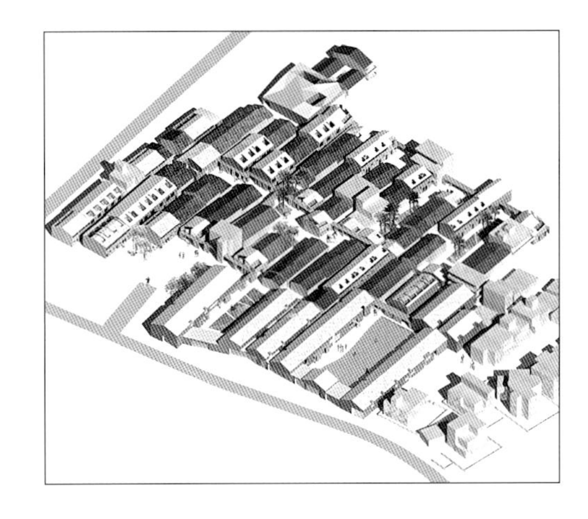

南油购物公园城市设计
Nanyou Shopping Park

地点 / Location：深圳 / Shenzhen
设计 / Design：2008
用地面积 / Site Area：130000m²
建筑面积 / Floor Area：455180m²
项目组 / Team：孟岩 | 黄志毅 吴文一 | 潘志国 饶恩辰 周亮 于金翠 黄艺宏 张震
业主 / Client：凯德置地

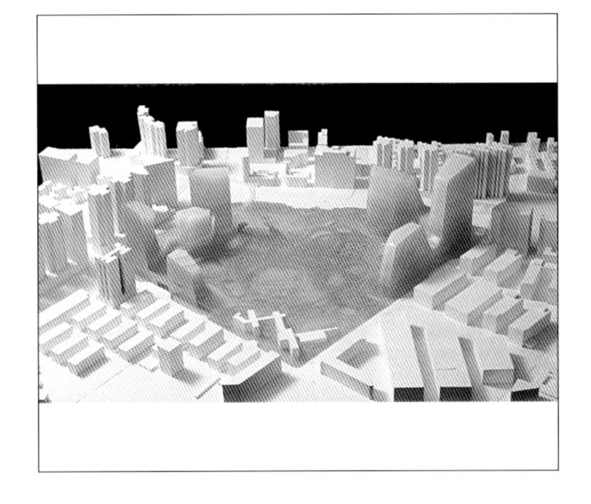

深圳公明中心区城市品质提升
Gongming Town Center Urban Improvement Study

地点 / Location：深圳 / Shenzhen
设计 / Design：2009
用地面积 / Site Area：900000m²
项目组 / Team：孟岩 | 王衍 | 欧阳祎 黄艺宏
合作 / Collaborators：香港中文大学廖维武
北京大学国土资源研究所 公众力商务咨询有限公司
业主 / Client：深圳市规划局光明分局

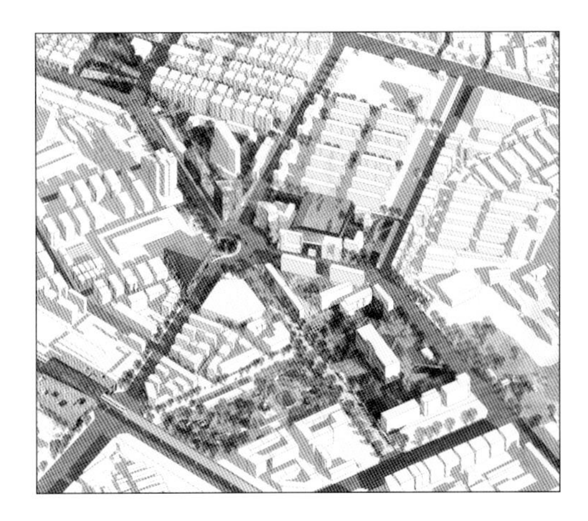

鄂尔多斯 CBD 概念规划设计
Ordos CBD Master Plan

地点 / Location：鄂尔多斯 / Ordos
设计 / Design：2010 ~ 2011
用地面积 / Site Area：274123m²
建筑面积 / Floor Area：1033279m²
项目组 / Team：刘晓都 | Manuel Sanchez-Vera 苏爱迪 | Michael H. Rogers
Mares E.C. Marc Alessandra Deidda
业主 / Client：内蒙古鄂尔多斯投资控股集团有限公司

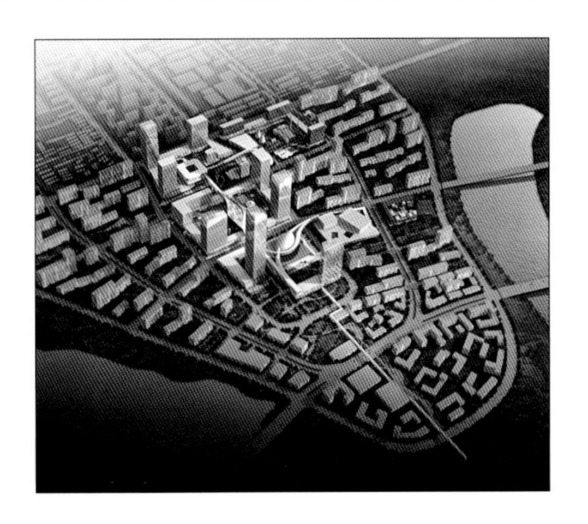

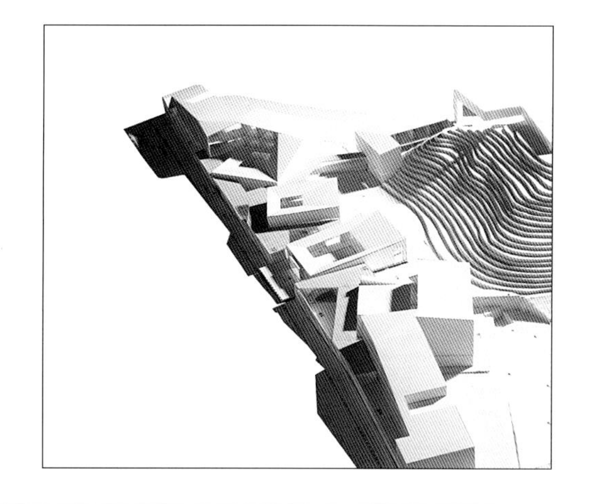

巽寮设计酒店
Design Hotel in Xunliao

地点 / Location：广东惠州 / Huizhou, Guangdong
设计 / Design：2009
用地面积 / Site Area：23243m^2
建筑面积 / Floor Area：12108m^2
项目组 / Team：刘晓都 | 王俊 刘明 | 李嘉嘉 何慧珊 胡志高
合作 / Collaborators：杨邦胜酒店室内设计（室内）
业主 / Client：惠东县鸿康实业有限公司

大连海·中国体育会所
Dalian Maritime Sports Club

地点 / Location：大连 / Dalian
设计 / Design：2008 ~ 2009
建成 / Completion：2011
用地面积 / Site Area：6000m^2
建筑面积 / Floor Area：5000m^2
项目组 / Team：王辉 | 吴文一 | 赵洪言 | 郝钢 孔祥栋 刘爽 陈春 杜爱宏 段云龙
合作 / Collaborators：大连都市发展建筑设计（施工图）
业主 / Client：华润（大连）房地产有限公司

太原万科品牌中心
Display Center for Vanke Real Estate Company

地点 / Location：太原 / Taiyuan
设计 / Design：2010
建成 / Completion：2011
用地面积 / Site Area：5733m^2
建筑面积 / Floor Area：2000m^2
项目组 / Team：王辉 | 郝钢 | 杜爱宏 孔祥磊 张永建 骆丽贤 索琪
合作 / Collaborators：太原市建筑设计研究院（施工图）
业主 / Client：北京万科企业有限公司

诸暨酒店
Zhuji Hotel

地点 / Location：浙江诸暨 / Zhuji, Zhejiang
设计 / Design：2008
用地面积 / Site Area：12050m^2
建筑面积 / Floor Area：51500m^2
项目组 / Team：刘晓都 | 陶剑坤 | 饶恩辰 夏淼 黄志毅 于金翠
业主 / Client：诸暨市旧城改造管理委员会办公室

泰州华侨城水岸商业街
Taizhou OCT Canal Shopping Street

地点 / Location：江苏泰州 / Taizhou, Jiangsu
设计 / Design：2007 ~ 2008
建成 / Completion：2011
用地面积 / Site Area：70000m²
建筑面积 / Floor Area：20000m²
项目组 / Team：王辉 | 刘旭 杜爱宏 | 王菁菁 段云龙 王鹏 Joseph Kan
张晓 熊星 马海东 张柳娟 陈通 张淼
合作 / Collaborators：江苏省建筑设计研究院（施工图）
业主 / Client：泰州华侨城投资发展有限公司

杭州西溪国家湿地艺术村酒店
Artist Community Hotel in Xixi National Wetland

地点 / Location：浙江杭州 / Hangzhou, Zhejiang
设计 / Design：2007 ~ 2010
建成 / Completion：在建 / Under Construction
用地面积 / Site Area：20000m²
建筑面积 / Floor Area：7286m²
项目组 / Team：刘晓都 | 黄志毅 吴春英 | 熊慧晶 何慧珊 于金翠
魏志姣 欧阳祎 李强 姚晓微 陈兰生
合作 / Collaborators：广州容柏生建筑工程设计深圳事务所（结构）
天宇机电设计（机电设备） 华优建筑设计院（施工图）
业主 / Client：杭州市余杭区建设局

深圳香蜜湖康体运动中心
Xiangmi Lake Golf Club

地点 / Location：深圳 / Shenzhen
设计 / Design：2006 ~ 2011
建成 / Completion：2011
用地面积 / Site Area：14715m²
建筑面积 / Floor Area：7078m²
项目组 / Team：刘晓都 | 张新峰 杨期力 | 夏淼 张星 嵇羽宇
Richie Gelles 傅卓恒 王雅娟 孙艳花 李强 黄煦
合作 / Collaborators：北京中外建建筑设计深圳分公司（施工图）
业主 / Client：深圳市宏威投资发展有限公司

南山区东滨路改造
The Streetscape Renovation of Dongbin Road

地点 / Location：深圳 / Shenzhen
设计 / Design：2007 ~ 2009
建成 / Completion：2009
用地面积 / Site Area：两公里的街景整治
项目组 / Team：刘晓都 吴文一 | 朱加林 | 左雷 | 程昀 陈耀光 乔锴
魏志姣 廖志雄 熊慧晶 黄艺宏 饶恩辰 黄煦
合作 / Collaborators：北方 - 汉沙杨建筑工程设计（施工图）
业主 / Client：深圳市南山区建筑工务局

深圳三洲田东部华侨城别墅 3# 地块
3# Villa of OCT EAST

地点 / Location：深圳 / Shenzhen
设计 / Design：2007 ~ 2008
用地面积 / Site Area：268267m²
建筑面积 / Floor Area：48390m²
项目组 / Team：孟岩 刘晓都 | 陶剑坤 | 黄志毅 杨期力 Mathias Wolf
姚晓微 程昀 乔锴 丁钰 涂江 钱进 周建杰
业主 / Client：深圳华侨城房地产有限公司

深圳大梅沙心海蓝桥酒店公寓
Daimeisha Xinhai Lanqiao Hotel & Apartment

地点 / Location：深圳 / Shenzhen
设计 / Design：2007 ~ 2010
用地面积 / Site Area：8148m²
建筑面积 / Floor Area：35816m²
项目组 / Team：孟岩 | 黄志毅 | 邓丹 姚晓微 左雷 Cedric Yu
欧阳祎 王俊 熊慧晶 李立德 孙艳花
合作 / Collaborators：同济人建筑设计（施工图）
业主 / Client：深圳市心海投资发展有限公司

深圳湾厦海境界一期
Wanxia Sea Realm Phase I

地点 / Location：深圳 / Shenzhen
设计 / Design：2007 ~ 2010
建成 / Completion：在建 / Under Construction
用地面积 / Site Area：12892m²
建筑面积 / Floor Area：85241m²
项目组 / Team：刘晓都 李雅丽 | 姚晓微 | 张新峰 傅卓恒 杨光 艾芸 刘明
饶恩辰 左雷 王俊 李耀宗 何慧珊 林怡琳 张英 熊慧晶 陈兰生
合作 / Collaborators：朗联设计（室内）华森建筑与工程设计（施工图）
业主 / Client：深圳市蛇口湾厦实业股份有限公司

汀风渡别墅
Wuxi Courtyard Townhouse

地点 / Location：无锡 / Wuxi
设计 / Design：2009
建成 / Completion：2011
用地面积 / Site Area：132600m²
建筑面积 / Floor Area：65000m²
项目组 / Team：王辉 吴文一 | 李淳 | 王菁菁 张永建 索琪 李湃 张柳娟
合作 / Collaborators：无锡市民用建筑设计院（施工图）
业主 / Client：阳光壹佰置业集团有限公司

无锡阳光 100 商业街
Sunshine 100 Shopping Center

地点 / Location：无锡 / Wuxi
设计 / Design：2008 ~ 2009
建成 / Completion：2011
用地面积 / Site Area：10000m²
建筑面积 / Floor Area：8170m²
项目组 / Team：王辉 | 吴文一 | 李淳 | 杜爱宏 郑娜
王菁菁 张淼 胡美铃 张柳娟 张耀辉 李湃
合作 / Collaborators：无锡市民用建筑设计（施工图）
业主 / Client：阳光壹佰置业集团有限公司

天津塘沽老年公寓
Tanggu Elderly Home

地点 / Location：天津 / Tianjin
设计 / Design：2008
用地面积 / Site Area：15900m²
建筑面积 / Floor Area：39100m²
项目组 / Team：王辉 吴文一 | 王菁菁 | 孔祥磊 郑娜 张耀辉 霍振舟 索琪 李湃
业主 / Client：天津市塘沽区民政局

南方科技大学教师及专家公寓
Faculty's Apartment for South University of Science and Technology

地点 / Location：深圳 / Shenzhen
设计 / Design：2010
建成 / Completion：在建 / Under Construction
用地面积 / Site Area：20498m²
建筑面积 / Floor Area：67037m²
项目组 / Team：孟岩 | 林怡琳 姚晓微 | 饶恩辰 黄志毅 Manuel Sanchez-Vera
周成松 许小东 王俊 姜轻舟 李耀宗 Mares E.C. Marc 邓军
合作 / Collaborators：深圳市建筑科学研究院（施工图） 互联盟建筑设计（施工图）
业主 / Client：深圳市建筑工务署 南方科技大学建设办公室

开心蚁居 - 万科回龙观极小住宅青年公寓
Happy Ant Farm - Housing with a Mission

地点 / Location：北京 / Beijing
设计 / Design：2011 ~ 2012
用地面积 / Site Area：1000m²
建筑面积 / Floor Area：3273m²
项目组 / Team：王辉 吴文一 | Jennife H. Ha 陈睿
业主 / Client：北京万科企业有限公司

都市实践简介

由刘晓都、孟岩和王辉创建于1999年，现已成长为一个具有国际影响力的建筑师团队。目前在北京、深圳及香港设有分支机构。本书呈现的二十余个作品选自2007年至今的设计项目。从中可看到URBANUS都市实践自始至终在坚持和实践它的理念：从广阔的城市视角和特定的城市体验中解读建筑的内涵，紧扣中国的城市现实，以研究不断涌现的当下城市问题为基础，致力于建筑学领域的探索，力图为新世纪建筑和城市所面临的新问题提供新的解决办法。建成作品如大芬美术馆、唐山城市展览馆及广州土楼社会住宅为都市实践赢得了国际声誉。事务所多次应邀参加国际知名建筑展览及交流活动，多项设计作品发表在中国《时代建筑》、《世界建筑》、《建筑学报》，意大利 *Domus*、*Abitare*、荷兰 *Mark*、*Frame*、美国 *Architectural Record*、英国 *Blueprint*、*a+d*，日本 *GA*、*a+u*、西班牙 *a+t*，德国 *Bauwelt* 等权威设计杂志，并多次获得重要的设计奖项。

URBANUS都市实践正努力拓展新的设计和研究平台，积极探讨更广泛的国际化跨领域合作模式。于2010年成立的研究部（URB）作为连接实践与学术的平台，探索由学术研究出发，对接实践设计的模式；梧桐山工作坊亦将于2012年内成立，力图使事务所在第二个十年的发展中迈向更高的台阶。

展览

2012 GA International 2012 展，东京

2012 从北京到伦敦——当代中国建筑展，伦敦

2011 向东方——中国建筑景观展，MAXXI 建筑博物馆，罗马

2011 ~ 2005 历届"深圳•香港城市/建筑双城双年展"，深圳、香港

2011 Daring Design, 荷兰建筑师协会（NAI），鹿特丹

2011 北京设计三年展，国家博物馆，北京

2011 "物我之境：田园/建筑/城市"国际建筑展，成都

2010 上海世博会深圳案例馆，总策展及展览设计，上海

2010 东西南北中－十年的三个民间叙事，北京、上海、成都、烟台

2010 设计的立场－中荷跨界设计展，北京，上海，成都，深圳

2009 ~ 2010 活的中国园林展，欧罗巴利亚中国文化年，布鲁塞尔

2009 ~ 2010 心造——中国当代建筑前沿展，欧罗巴利亚中国文化年，布鲁塞尔

2008 ~ 2009 土楼公舍——中国廉租住宅个展，库珀.休伊特国家设计博物馆，纽约

2008 中国建筑五人展，纽约建筑中心

2008 当代中国建筑展，法国建筑师学会，巴黎，巴塞罗那

2007 里斯本国际建筑与设计三年展，葡萄牙建筑师协会，世博馆，里斯本

2006 "当代中国"建筑、艺术与视觉文化展，荷兰建筑师协会（NAI），鹿特丹

2005 圣保罗国际建筑与设计双年展，巴西建筑师协会，圣保罗

2005 第二届广州当代艺术三年展，广州

获奖

2012 第四届美国《建筑实录》最佳居住建筑设计奖及最佳改造设计奖

2011 UED 博物馆建筑设计奖

2010 阿卡汗奖提名

2010 第三届美国《建筑实录》中国奖2项最佳公共建筑奖及年度建筑大奖

2008 中国建筑传媒奖之"居住建筑特别奖"

2008 第四届 WA《世界建筑》奖优胜奖和佳作奖

2008 第二届美国《商业周刊/建筑实录》中国奖"室内设计奖"和"公共建筑奖"

2007 T+A 建筑中国"年度建筑设计机构奖"

2007 深圳•香港城市/建筑双城双年展"最佳公众奖"

2006 首届美国《商业周刊/建筑实录》中国奖"最佳公共建筑奖"

2005 美国《建筑实录》年度全球设计先锋

2004 第二届 WA《世界建筑》佳作奖和鼓励奖

URBANUS Profile

Founded in 1999, under the leadership of partners Xiaodu Liu, Yan Meng and Hui Wang, URBANUS developed its branches in Beijing, Shenzhen and Hong Kong. It is recognized as one of the most influential architecture practices in China. Over 20 projects of URBANUS, dated from 2007 to present, are selected in this book. All of them reaffirm URBANUS' core values and constantly and consciously put this value into practice: interpreting the internal meaning of architecture with the macroscopic perspective of urbanism and the specific urban context; actively confronting to the reality of contemporary urban China and exploring novel territory of architecture based on in-depth research about urban problems; and proposing novel solutions to the problems of the next generation architectural and urban designs. Many of URBANUS' completed projects, such as Tulou Collective Housing, Dafen Art Museum and Tangshan Urban Planning Museum, achieved an international reputation for the office. The office was also invited to participate in many internationally prominent exhibitions and events. Many projects of URBANUS were awarded and widely published in mainstream architecture magazines and journals, such as *T+A, W+A, Domus, Abitare, Mark, Architecture Record, Blueprint, a + u, a + t, Bauwelt,* etc..

URBANUS is now developing and extending its existing design and research platform, and exploring opportunities of international and multidisciplinary collaborations. Urbanus Research Bureau (URB) was established in 2010. This research unit aims to develop the platform for interactions between profesional practice and academia, and to explore the model of research-oriented design practice. Wutongshan Workshop will be launching in 2012. These together will set the critical cornerstone for Urbanus' development in its second decade.

EXHIBITIONS

2012 GA International 2012, Tokyo.

2012 From Beijing to London, 16 Contemporary Chinese Architects, London

2011 VERSO EST Chinese Architectural Landscape, Rome

2011 ~ 2005 "The Shenzhen & Hong Kong Bi-City Biennial of Urbanism/Architecture", Shenzhen, Hong Kong

2011 "Daring Design", Rotterdam

2011 "Beijing Design Triennial", China National Museum, Beijing

2011 "Holistic Realm: Landscape / Architecture / Urbanism" Architectural Biennale Exhibition Space and Preliminary layout, Chengdu

2010 Shanghai Ex-po, Shenzhen Case Pavilion, General Curator and Exhibition Designer

2010 "Compass" – Three Narrative about the Past Ten Years, Beijing / Shanghai / Chengdu / Yantai

2010 "Taking a Stance-8 Positions", Beijing / Shanghai / Chengdu / Shenzhen

2009 ~ 2010 "Chinese Gardens for Living, Europalia 2009", The Square, Belgium

2009 ~ 2010 "Building from Heart, Europalia 2009", The Square, Belgium

2008 ~ 2009 "Solos-Tulou: Affordable Housing for China", Cooper Hewitt National Design Museum, New York

2008 "Building China", Architecture Center, New York

2008 "Exhibition of Chinese Contemporary Architecture", Paris / Barcelona

2007 "Lisbon Architecture Triennial", Portugal Pavilion, Lisbon

2006 "China Contemporary" Architecture, Art, Visual Culture, Rotterdam

2005 "The 26th Sao Paulo Biennial", Sao Paulo

2005 "The 2nd Guangzhou Contemporary Art Triennial", Guangzhou

AWARDS

2012 Best Residencial Project and Best Preservation Project, Architectural Record China Awards 2012

2011 UED Museum Architecture Design

2010 Nomination of Aga Khan Award for Architecture

2010 Best Public Project, Architectural Record China Awards 2010 (two projects) and Project of the Year

2008 China Architecture Media Award, Residential Architecture Special Prize

2008 WA Chinese Architecture Award "Winning Prize" & "Honorable Mention"

2008 Best Public Project /Best Interior Project, Business Week /Architectural Record China Awards

2007 T+A Annual Prize of Architectural Design Organization

2007 Shenzhen & Hong Kong Bi-City Biennial of Urbanism /Architecture-Best Public Prize

2006 Business Week /Architectural Record China Awards - Best Public Project

2005 Architectural Record / Design Vanguard

2004 WA Chinese Architecture Award "Honorable Mention" & "Merit Award"

名录

Staff List

1999 ～ 2007（详《URBANUS 都市实践》2007）

2008

刘晓都 孟岩 王辉 |（深圳）陈春 邢果 张慧文 朱加林 傅卓恒 林海滨 黄志毅 王彦峰 涂江 刘子荣 黄煦 廖素珍 姚晓微 陶剑坤 左雷 彭宇新 夏淼 尹毓俊 何珂斯 张云 李昊 戴福平 林怡琳 任嘉利 王衍 饶恩辰 魏志姣 韩磊 艾芸 何慧珊 张新峰 席江 张明慧 于金翠 杨光 李强 廖志雄 吴贞 孙艳花 熊慧晶 王俊 张震 周亮 潘志国 胡志高 熊嘉伟 裘石 黄艺宏 欧阳祎 吴春英 李立德 龚潇婧 何勇（北京）魏燕 陈春 段云龙 朱庆会 张永清 张淼 杜爱宏 张永建 刘爽 郝钢 赵洪言 陈迪 张柳娟 贾鸳鸳 林秀清 陈岚 孔祥磊 索琪 李淳 王菁菁 成直 陈通 张耀辉 杨勍 郑娜 吴文一 李湃 唐康硕 霍振舟 孔祥栋 李明光 张晓兰 程清 王鹏 谢金山 江海洋 黄灿辉 张兴斌

2009

刘晓都 孟岩 王辉 |（深圳）陈春 张慧文 朱加林 傅卓恒 林海滨 黄志毅 王彦峰 黄煦 廖素珍 姚晓微 左雷 彭宇新 何珂斯 张云 李昊 戴福平 林怡琳 任嘉利 王衍 饶恩辰 魏志姣 韩磊 艾芸 何慧珊 张新峰 席江 杨光 李强 廖志雄 吴贞 孙艳花 熊慧晶 王俊 张震 潘志国 胡志高 熊嘉伟 裘石 黄艺宏 欧阳祎 吴春英 李立德 何勇 胡伊硕 梁广发 刘明 罗仁钦 张丽娜 陈兰生 黄俊 苏爱迪 黄中汉 张丹丹 李嘉嘉 林挺 邢月 刘思钊（北京）魏燕 陈春 段云龙 朱庆会 张永清 张淼 杜爱宏 张永建 刘爽 郝钢 赵洪言 陈迪 贾鸳鸳 林秀清 孔祥磊 李淳 王菁菁 成直 杨勍 郑娜 吴文一 唐康硕 姚东杰 曾皓 杨楠 李佳颖 张晓兰 刘银燕 张柳娟 陈岚 陈通 张耀辉 霍振舟 孔祥栋 李湃 索琪 包瑞 崔海峰

2010

刘晓都 孟岩 王辉 |（深圳）陈春 张慧文 朱加林 林海滨 黄志毅 王彦峰 廖素珍 姚晓微 左雷 彭宇新 何珂斯 张云 戴福平 林怡琳 王衍 饶恩辰 魏志姣 艾芸 张新峰 席江 李强 廖志雄 孙艳花 熊慧晶 王俊 张震 胡志高 裘石 黄艺宏 欧阳祎 吴春英 李立德 何勇 胡伊硕 梁广发 刘明 罗仁钦 张丽娜 陈兰生 苏爱迪 黄中汉 张丹丹 李嘉嘉 林挺 邢月 刘思钊 祝玉梅 叶蕾 周娅琳 邓军 温谨馨 吴晓丽 林恬 李莎 曾冠生 Deidda Alessandra 郭静 Michael H. Rogers Cressica J. Brazier Manuel Sanchez-vera Gomez-trelles 张天欣 臧敏 Mares E. C. Marc 朱伶俐 张长文 李耀宗 周成松 吴先群（北京）魏燕 陈春 段云龙 朱庆会 张淼 杜爱宏 张永建 刘爽 郝钢 赵洪言 陈迪 贾鸳鸳 林秀清 孔祥磊 李淳 王菁菁 成直 郑娜 吴文一 唐康硕 姚东杰 王韬 杨涛 刘妮妮 郑明路 Jennifer H.Ha Deborah A. Richards 徐晶 陈睿 张晓兰 曾皓 杨楠 李佳颖 张广瀚 张永清 杨勍

2011

刘晓都 孟岩 王辉 |（深圳）陈春 张慧文 朱加林 林海滨 黄志毅 王彦峰 廖素珍 姚晓微 彭宇新 何珂斯 张云 戴福平 林怡琳 饶恩辰 魏志姣 艾芸 张新峰 李强 孙艳花 王俊 张震 裘石 黄艺宏 罗仁钦 陈兰生 苏爱迪 黄中汉 张丹丹 李嘉嘉 林挺 邢月 刘思钊 叶蕾 周娅琳 邓军 温谨馨 吴晓丽 林恬 李莎 Michael H. Rogers Cressica J. Brazier 张天欣 臧敏 朱伶俐 张长文 李耀宗 周成松 吴先群 沈振中 王鹏 林达 傅娜 刘子荣 孔倩珺 张海君 Ben Wu 于晓兰 陈丹平 许小东 谢盛奋 姜轻舟 刘洁 吴锦彬 张英 潘远芳 徐罗以 姚殿斌 陈文赟 王亦乐 陈婧波 廖俊 吴姿桦 赵凤伟（北京）魏燕 陈春 杜爱宏 张永建 刘爽 郝钢 赵洪言 陈迪 林秀清 吴文一 孔祥磊 王菁菁 成直 郑娜 王韬 刘妮妮 郑明路 Jennifer H.Ha 徐晶 陈睿 Samuel T. Ruberti 张晓兰 段云龙 朱庆会 张淼 贾鸳鸳 李淳 唐康硕 姚东杰 杨涛 Deborah A. Richards David B. Hooks Patrick O. Candalla（香港）Chritso Logan Danil Nagy Silan Yip 纪逸纯 纪家溢 陈德扬 徐启豪

2012

刘晓都 孟岩 王辉 |（深圳）陈春 张慧文 林海滨 王彦峰 廖素珍 姚晓微 彭宇新 何珂斯 张云 戴福平 林怡琳 饶恩辰 魏志姣 张新峰 孙艳花 王俊 裘石 罗仁钦 陈兰生 黄中汉 林挺 刘思钊 叶蕾 周娅琳 温谨馨 吴晓丽 李莎 Michael H. Rogers 张天欣 臧敏 朱伶俐 张长文 李耀宗 周成松 吴先群 沈振中 林达 傅娜 刘子荣 张海君 于晓兰 陈丹平 许小东 谢盛奋 姜轻舟 刘洁 吴锦彬 张英 潘远芳 徐罗以 姚殿斌 陈文赟 陈婧波 廖俊 吴姿桦 赵凤伟 余欣婷 胡润泽 韩潇 吴晗 姚少玲 赵佳 袁柏浩 何志花 冯绚 苏彦 谭秀敏 王燕萍 刘羽 吴汉成 吴然 Anna Laura Govoni Camilla Costa Cristina Peraino（北京）魏燕 陈春 杜爱宏 张永建 刘爽 郝钢 赵洪言 陈迪 林秀清 吴文一 孔祥磊 王菁菁 成直 郑娜 王韬 刘妮妮 郑明璐 Jennifer H.Ha 徐晶 陈睿 Samuel T Ruberti 王立杰 苗艳清 刘聪 李图 骆丽贤 金晨佳 杨柳 李强 Suliciu Ioana 王亦乐 王求安 叶宁（香港）Chritso Logan Danil Nagy Silan Yip 纪逸纯 纪家溢 陈德扬 Travis Bunt Stephanie Youngji Choi 邬程光 罗美坚 黄利鈞

（实习生）

张烁 林菊莲 庞琨伦 何水 洪涵 孔繁锦 刘声 黄勇 张文青 齐帆 陈成发 肖杨 陈志威 李文烜 李小燕 金泰山 郑巨信 王茜 陈颂威 王欣 孟成 高浩 汪澄波 周璐菡 汪源 王晶晶 谢洁 孟浩 张雯燕 蔡勉 马晓瑛 陈翔融 罗咏恒 罗颖思 陈斯 陈忱 冯立星 李泳贤 钟铖 吴少文 常辰 周月 马宁 吴妍玥 张丽丽 吴迪 吴翌 吴凡 陈瑜 王欣峰 胡雅琳 张叶 李莉 张力 黎恢远 张文涛 俞乔 刘兵 张雅楠 阴汀 沈若禹 常俊颖 鲁杨 李憬君 杨益 夏黎 肖伟明 李林 赵玮 曾安培 黄晗 郎杏枝 郑汉明 罗俊辉 孙佳 王峥 李真百 王芳 王燕萍 李蕊 刘羽 谢文珊 王瑛子 王越 甄琪 李明基 贾砚琦 余雪婧 张雪石 王寅 张轶伟 汤迪莎 罗婧 温茜玥 孙冠男 王祥笛 孟海洋 刘心如 雷沐羲 韩如意 麦卓恒 刘健涛 龙萧合 雍玉洁 苏武顺 唐州 全水 崔一松 万妍 曹梦然 钟鸿毅 劳康勇 郭冉 郭旭生 林煜涛 梁靖 徐徜徉 张晓 李喆 张鑫 郑晨亮 董鹏 杨永宽 崔轩 江可唯 曾皓 刁愷 杨健 戴天行 罗晶 裴雷 张书勤 胡美玲 Kan Joseph Rautio S.Maria 吴一凡 吴浩 徐少楠 张园华 张璐 饶岗 彭喆 马喆 吴诗静 张哲 张晓鸣 王晓雨 夏海玲 邓思颖 王茜 王朗 栾博 周硕安 郭凡 陈宣儒 高大智 武州 王颖 吴志杰 陶一兰 王川 沈一婷 尹文刚 丁伟斌 郭菁 祁伟 王莹 米键 闵一洋 曹凯中 王悦 张柯达 廖翕 王宇瞳 王宇 耿晓蕾 陈铎 尤伟阳 张超 伍毅敏 Daniel Ladyman 冯雨晴 皮天祥 吕亚辉 陈龙 崔巍懿 徐尧 张颖 王耀龙 David Wilson